FINAL FRONTIER

Canada

VOYAGES INTO OUTER SPACE

DAVID OWEN

FIREFLY BOOKS

A FIREFLY BOOK

Published by Firefly Books Ltd., 2004

First printing

Publisher in Cataloguing-in-Publication Data (U.S.)

Owen, David.
Final frontier : voyages into outer space / David Owen. – 1st ed.
[128] p. ; col. photos. : cm.
Includes index.
Summary: Space exploration from early attempts to the future including: the Apollo missions, space shuttle, international space station, Hubble Space Telescope and unmanned space probes.
ISBN 1-55297-776-5
ISBN 1-55297-775-7 (pbk.)
1. Outer space—Exploration. 2. Astronautics—Popular works. I. Title.
629.4 21 TL793.O971 2004

National Library of Canada Cataloguing in Publication Data

Owen, David
Final frontier : voyages into outer space / David Owen.
Includes index.
ISBN 1-55297-776-5 (bound).—ISBN 1-55297-775-7 (pbk.)
1. Outer space—Exploration—Juvenile literature. I. Title.
TL793.O94 2004 j629.4 C2003-900785-5

Published in Canada in 2004 by
Firefly Books Ltd.
66 Leek Crescent
Richmond Hill, Ontario L4B 1H1

Published in the United States in 2004 by
Firefly Books (U.S.) Inc.
P.O. Box 1338, Ellicott Station
Buffalo, New York 14205

This book was designed and produced by
Quintet Publishing Limited
6 Blundell Street
London N7 7BH

FFRO

Text adaptation: Victoria Sherrow
Editor: Catherine Osborne
Designer: James Lawrence

Creative Director: Richard Dewing
Publisher: Oliver Salzmann

Manufactured in Singapore by Universal Graphics Pte Ltd
Printed in China by Midas Printing International Ltd

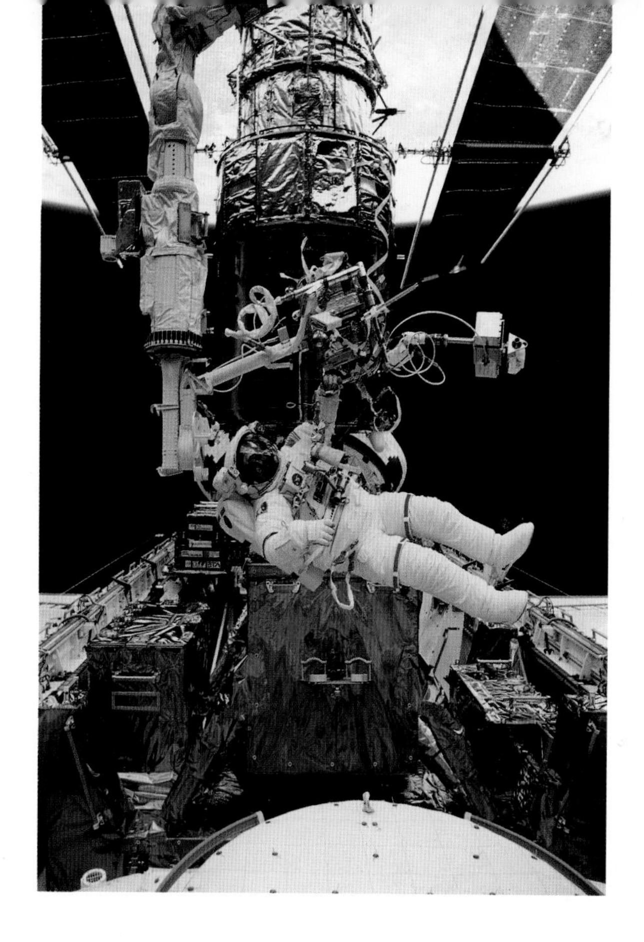

Contents

1

THE MAPPING OF SPACE

Step by step, pioneer astronomers gained new knowledge of the heavens, which later enabled people to launch spacecraft to distant planets and send humans to the Moon.

Since humans first walked upon the Earth, they have been awed by the spectacle of the starlit sky. Ancient civilizations named groups of stars after their gods or heroes. The Babylonians used calculations to predict planetary movements, and the Mayans of Mexico and Central America studied the stars to determine planting and harvesting times. The ancient monument of Stonehenge in England may be a complex astronomical computer.

Some 2,000 years ago, astronomy began to develop as a science. One of the pioneers was Claudius Ptolemy, a Greek working in Egypt around AD 127–41. Using his naked eye, Ptolemy mapped the heavens. He thought the Earth was stationary, with the Sun, the Moon and the planets all revolving around it on circular orbits. Centuries later, Ptolemy's basic ideas would be proven wrong.

In the meantime, humans found new ways to use astronomy. On clear nights, navigators at sea could spot true north by locating the Pole Star. Later, they used instruments, such as the sextant, and printed tables showing the positions of the stars to find their ship's position on an ocean voyage.

Ptolemy's map came under fire in 1543. A Polish priest and mathematician named Copernicus (or Niklas Koppernigk) noted variations in the way planets moved, which Ptolemy's ideas did not explain. He also thought that the Sun, not Earth, was

OPPOSITE An illustration of Ptolemy of Alexandria c AD 130, taken from Margarita Philosophica, 1535.

TOP Italian astronomer and physicist Galileo Galilei (1564–1642).

BELOW Woodcut of Copernicus (1473–1543).

ABOVE Tycho Brahe (1546–1601), taken from Tycho Brahe Opera, Vol IV, 1596.

at the center of the universe while Earth, and all the other planets, rotated in circular orbits around it.

That century brought another big discovery. In 1572, a new star appeared in the constellation of Cassiopeia, blazing so brightly that it could be seen in the daytime. This was a nova, although nobody knew this at the time. Yet its real importance lay in where it was. Using instruments created by a Danish astronomer, Tycho Brahe, astronomers realized that this new star must be a colossal distance from Earth.

Brahe thought that Copernicus's map was partly wrong. He said that all the planets but Earth moved around the Sun. Both the Sun and the Moon, said Brahe, were in orbit around Earth. But a young German mathematician and astronomer, Johannes Kepler, questioned this theory. Kepler became Brahe's assistant in 1600. After Brahe died in 1602, Kepler studied his paperwork and calculations. They did not match his observations of how the planets moved, especially Mars. Its orbit was measurably longer on one side of the Sun than it was on the other. By 1606, Kepler had concluded that the planets moved in elliptical orbits, not circular ones.

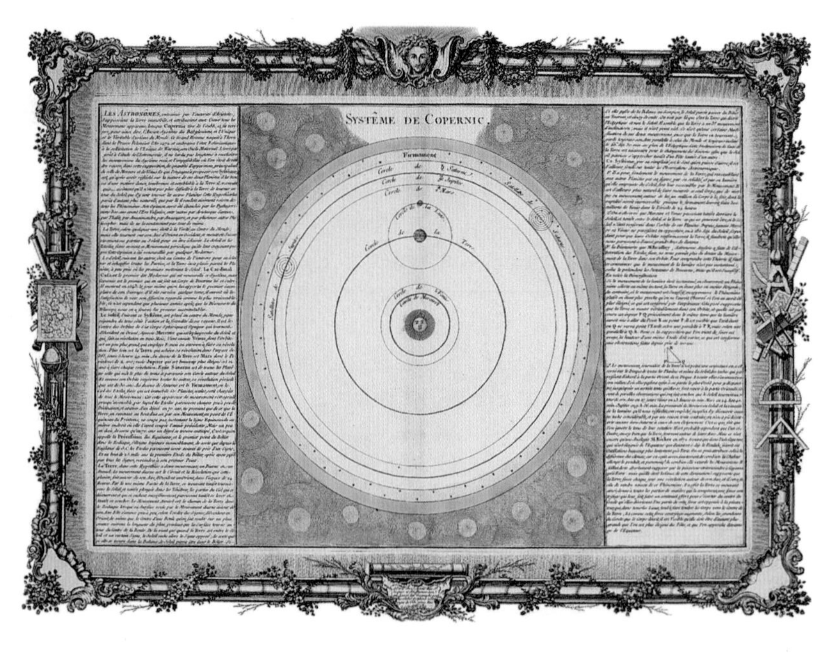

Progress was made with the invention of the telescope. The Dutch developed a type of spyglass, which they brought to the trading city of Venice, Italy, in 1609. A scientist and mathematician from Padua named Galileo Galilei was working there. After he heard about the new invention, he figured out how to make his own device, which magnified distant images about three times. Within months, he had made a telescope with a magnification factor of 30. Galileo explored the night skies and saw many previously invisible stars. He scanned the surface of the Moon and its craters and mountain ranges.

When he studied Jupiter, the largest planet of the solar system, he was surprised to find that four bodies which had looked like stars were actually planets or moons. They were in orbit around Jupiter, like the Moon circling Earth. Galileo realized that if Jupiter could have four moons and still orbit around the Sun, maybe the Earth and its one moon did the same? He went on to study Venus. His observations supported Copernicus and Kepler's explanation of the relationship of the planets, as they orbited around the Sun at different distances from it.

In 1610 and 1632, Galileo wrote two famous papers discussing his theories. They clashed with the teachings of the Catholic Church, which placed

ABOVE Johannes Kepler, (1571–1630).

OPPOSITE BOTTOM Hand-colored engraving showing Copernicus's planetary system, issued in Paris in 1761.

BELOW Sketches of the moon from Galileo's *Sidereus Nuncius*, 1610.

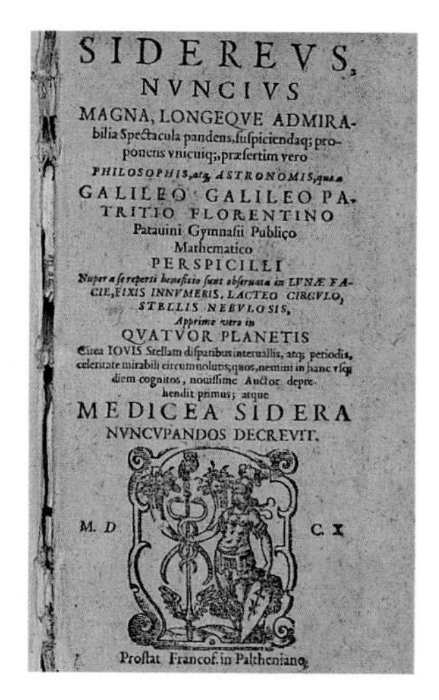

SIDEREVS,
NVNCIVS
MAGNA, LONGEQVE ADMIRAbilia Spectacula pandens, suspiciendaq; proponens vnicuiq;, præsertim vero
PHILOSOPHIS, atq; ASTRONOMIS, quæ
GALILEO GALILEO PATRITIO FLORENTINO
Patauini Gymnasii Publico Mathematico
PERSPICILLI
Nuper a se reperti beneficio sunt obseruata in LVNÆ FACIE, FIXIS INNVMERIS, LACTEO CIRCVLO, STELLIS NEBVLOSIS,
Apprime vero in
QVATVOR PLANETIS
Circa IOVIS Stellam disparibus interuallis, atq; periodis, celeritate mirabili circumuolutis; quos, nemini in hanc vsq; diem cognitos, nouissime Auctor deprehendit primus; atque
MEDICEA SIDERA
NVNCVPANDOS DECREVIT.
M. D C. X
Prostat Francof. in Paltheniano.

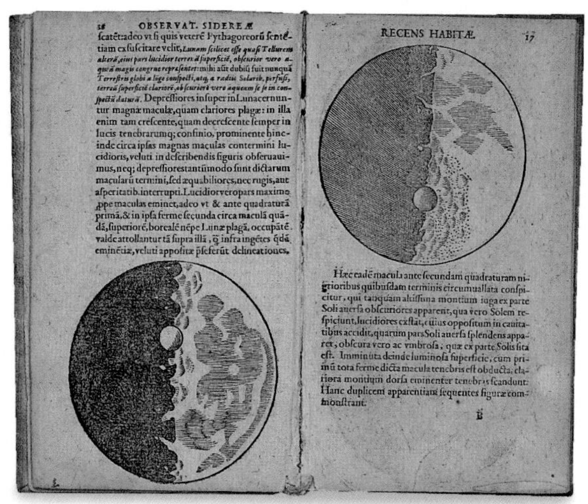

16 OBSERVAT. SIDEREÆ

scatēt: adeo vt si quis veterē Pythagoreorū sententiam exsuscitare velit, *Lunam scilicet esse quasi Tellurem alterā, eius pars lucidior terrenā superficiē, obscurior vero aqueā magis congrue repræsentet:* mihi aūt dubiū fuit nunquā *Terrestris globi a lōge conspecti, atq; a radiis Solarib. perfusi, terreā superficiē clariorē, obscuriorē vero aqueam se se in conspectū daturā.* Depressiores insuper in Luna cernuntur magnæ maculæ, quam clariores plagæ: in illa enim tam crescente, quam decrescente semper in lucis tenebrarumq; confinio, prominente hincinde circa ipsas magnas maculas contermini lucidioris, veluti in describendis figuris obseruauimus, neq; depressiores tantūmodo sunt dictarum macularū termini, sed æquabiliores, nec rugis, aut asperitatib. interrupti. Lucidior vero pars maxime ꝓpe maculas eminet, adeo vt & ante quadraturā primā, & in ipsa ferme secunda circa maculā quādā, superiorē, borealē nēpe Lunæ plagā, occupātē valde attollantur tā supra illā, q̄ infra ingētes q̄dā eminētiæ, veluti appositæ ꝓseferūt delineationes.

RECENS HABITÆ. 17

Hæc eadē macula ante secundam quadraturam nigrioribus quibusdam terminis circumuallata conspicitur, qui tanquam altissima montium iuga ex parte Soli auersa obscuriores apparent, qua vero Solem respiciunt, lucidiores exstāt, cuius oppositum in cauitatibus accidit, quarum pars Soli auersa splendens apparet, obscura vero ac vmbrosa, quæ ex parte Solis sita est. Imminuta deinde luminosa superficie, cum primū tota ferme dicta macula tenebris est obducta, clariora montium dorsa eminenter tenebras scandunt. Hanc duplicem apparentiam sequentes figuræ commonstrant.

B

Tools of the astronomer's trade — refracting and reflecting telescopes

Galileo and Kepler both used refracting telescopes, which used a double lens to produce magnified images. Galileo matched a concave lens with a convex magnifying lens in a tube. The first lens was the object lens, which focused light from distant planets onto the second lens, the eye lens, which magnified the resulting image. Kepler created an improved version with two convex magnifying lenses. His telescope had a wider field of vision and produced clearer images than Galileo's instrument.

These early refracting telescopes were limited because their lenses tended to break up white light into a spectrum of colors. Sir Isaac Newton designed a telescope that avoided this break up. He used a curved mirror to collect the light and deflect it into a magnifying eyepiece instead of a lens. This was the first reflecting telescope, and they soon dominated the field. Later, better-quality lenses made refracting telescopes more efficient. But reflecting telescopes remained supreme because they were relatively compact and it was easier to support very large mirrors than equally large lenses. As of 2002, the largest reflecting telescopes are the Mount Palomar telescope in California with a 200-inch (5 meter) diameter mirror, and the Zelenchukskaya telescope in the Caucasus with a 240-inch (6 meter) diameter mirror.

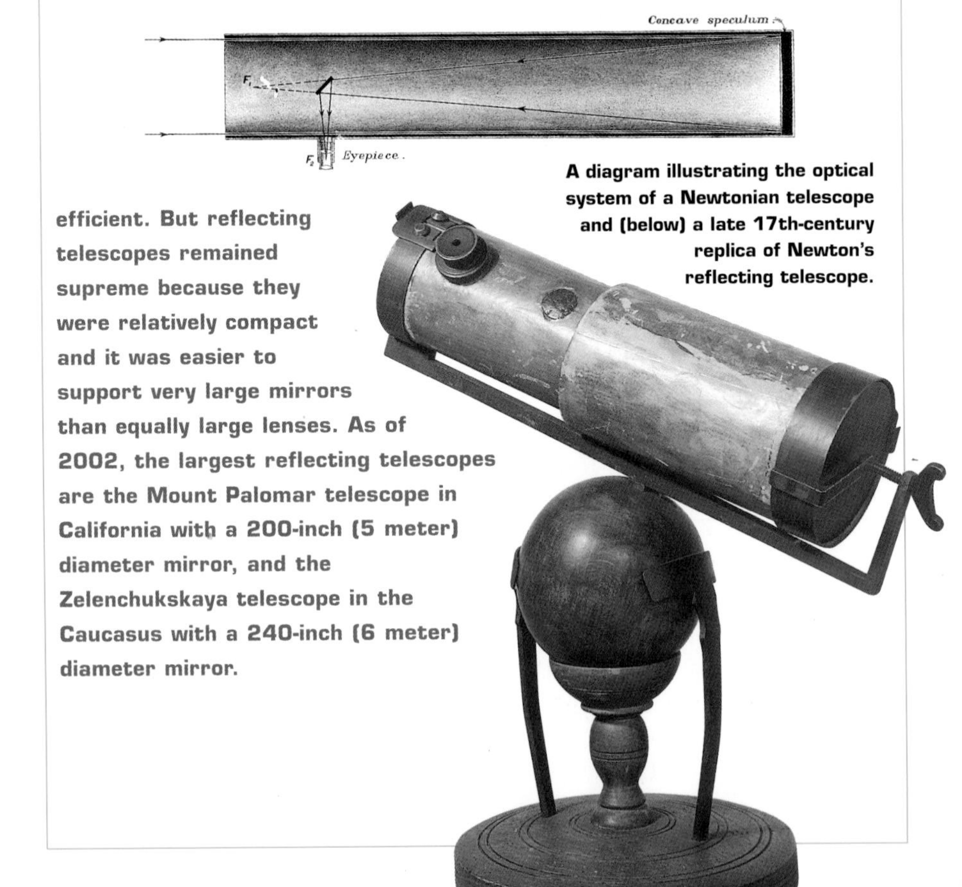

A diagram illustrating the optical system of a Newtonian telescope and (below) a late 17th-century replica of Newton's reflecting telescope.

Earth at the center of the solar system. Galileo was put on trial in Rome in 1633 and found guilty of heresy. Threatened with torture, he was forced to deny all his work, then placed under house arrest and forbidden to discuss or publish anything about astronomy. Galileo died nine years later, still confined to his house. For more than two centuries, all Catholics were forbidden to read his papers. But other scientists continued to ask questions and make observations that revealed the laws of space and how the universe works.

ABOVE Sir Isaac Newton (1642–1727).

Kepler had said that the length of a planet's orbit was proportional to its distance from the Sun. He suggested that some kind of force between Earth and the Moon maintained that orbit and caused tides to form. But how was it possible for Earth to move through space without this being obvious to anyone standing on its surface?

The answers came from a brilliant British mathematician, Sir Isaac Newton. Newton suggested that all objects attract other objects relative to their size. A person standing on the Earth's surface is held to Earth by gravity, but the person's mass also attracts Earth by an immeasurably small amount. Furthermore, gravity also holds the atmosphere in place, so there is no sensation of Earth's movement through space at its surface. Earth, the atmosphere and everything on its surface are all moving through space at exactly the same velocity. The only signs of movement are the Sun's path across the heavens by day, and the rotation of the stars about the Pole Star at night.

With his three Laws of Motion, Newton could explain all planetary movements. His first law stated that any immobile body will either stay at rest in the same position, or will continue its motion in a straight line, unless acted on by an outside force. The second law stated that the action of an outside force on a body will make it speed up in the direction of the force, by an amount proportional to the size of the force. Newton's third law went on to state that every action has an equal and opposite reaction.

TO THE FARTHEST REACHES:

observations by radio telescope

Optical telescopes use film to collect light over a long period, so they can reveal the presence of stars that would otherwise be invisible to humans. But the radio telescopes that began developing in the 1930s can "see" even farther. An American communications engineer, Karl Jansky, was searching for the source of noise on radio-telephone channels when he recorded radio waves coming to Earth from space. These radio waves cover a longer distance than does visible light, so they can reveal the presence of stars and galaxies far beyond the range of optical telescopes.

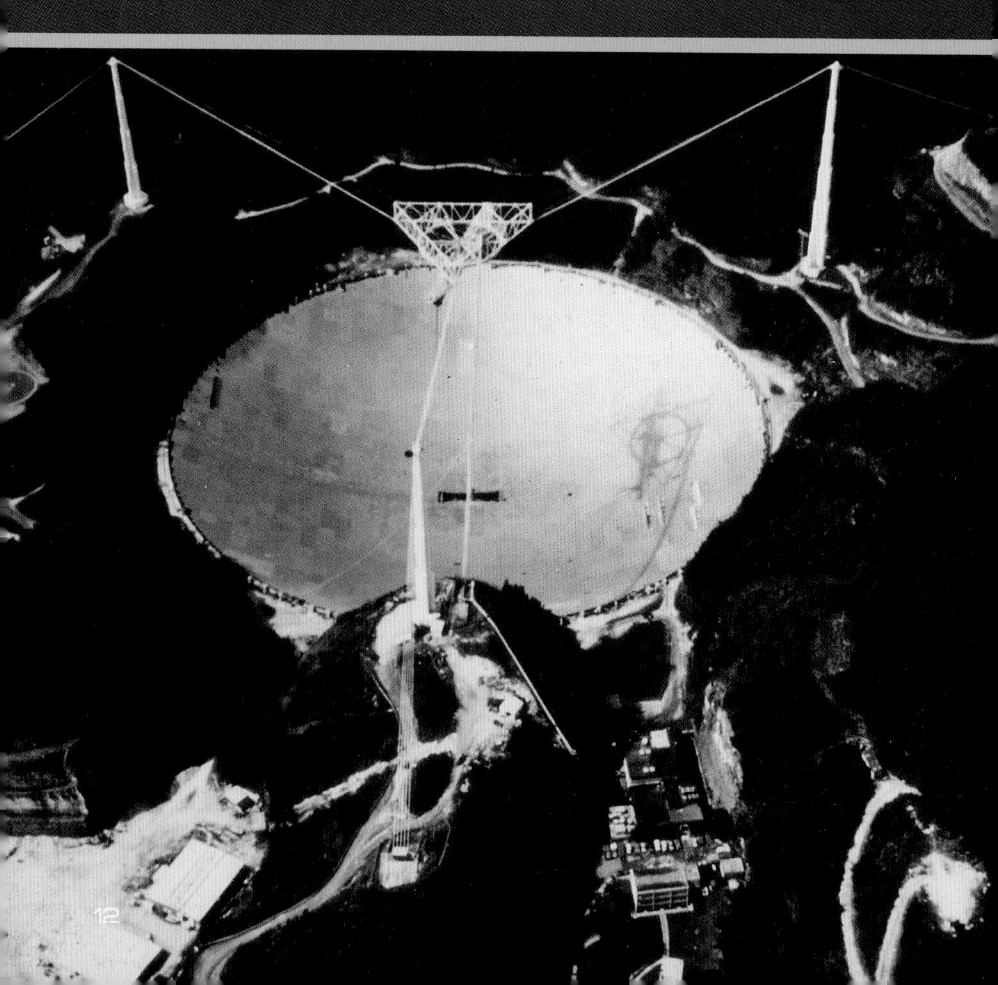

Radio telescopes once required very large "dishes" to help focus and record their long wavelengths. The largest single-dish radio telescope was built in the mountains of Puerto Rico in 1963, with a fixed dish 1,000 feet (305 meters) across. More recently, computers have enabled astronomers to combine signals from linked radio telescopes to make a picture equal to that of a much larger dish. The Very Large Array telescope in New Mexico, built in the early 1980s, produces an image equivalent to that generated by a dish 17 miles (27 kilometers) across.

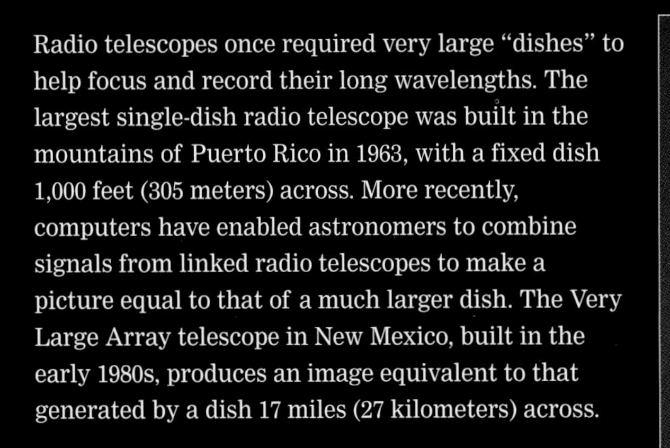

OPPOSITE Arecibo Observatory antenna at the National Astronomy and Ionosphere Center in Puerto Rico.

BELOW An observatory color density rendition view of the Sun.

ABOVE An orrery made in 1790, a mechanical device designed to reproduce the relative movements of Earth, Moon, Mercury and Venus around the Sun.

BELOW Spherical halo of Neutrinos around the Milky Way galaxy.

Newton proposed that a body attracts any other body with a force directly proportional to the result of their masses multiplied together, and inversely proportional to the square of the distance between them. This explained why the Moon stayed in orbit around Earth, for example. Newton's first law showed its natural tendency was to move in a straight line, but this was balanced by the gravity between Earth and the Moon that forced it to follow a regular orbit around the planet.

At last scientists could explain how and why the planets move around the Sun. Moreover, Newton's concept of gravity would eventually help scientists control a spacecraft so that it followed a desired path through the gravitational fields of different planets across the solar system and into deeper space. His findings, first published in 1682, have remained vital in space exploration.

Other scientists built on this work and added new ideas. Newton and others thought that the speed of light should vary depending on whether it was measured with or against the rotation of Earth. Early in the 20th century, however, German-born physicist Albert Einstein based his Theories of Relativity on the assumption that the speed of light never changes. In the "Special Theory of Relativity," he showed that because the speed of light is a constant, time would pass at different rates depending on the speed of the person measuring it. Similarly, mass and energy can be converted from one to the other. Einstein went on to develop the general theory of relativity, which showed that the effects of gravity and acceleration are identical.

A German astronomer, Friedrich Bessel, demonstrated the size of the universe beyond the solar system. He noticed an apparent change of position of a distant star, 61 Cygni, caused by Earth moving around the Sun. He calculated the distance between Earth and the star as approximately 60 million million miles (96 million million kilometers).

In 1851, a French physicist showed that Earth was rotating about its own axis. By then, improved telescopes had identified two more planets: Uranus (1781) and Neptune (1846). The outermost planet, Pluto, was finally spotted in 1930. William Herschel, who had first identified Uranus, also studied the densest pattern of stars in the night sky: the Milky Way. He suggested that the solar system is part of a much larger group of stars called a galaxy.

In 1918, American astronomer Harlow Shapley measured the distances of clusters of stars within the galaxy, and used his findings to calculate the size of the whole galaxy as 100,000 light-years across. Furthermore, the Sun was outclassed by powerful neighbors among the 100,000 million galaxy stars.

Six years later, another American, Edwin Hubble, confirmed the importance of the Milky Way. Using

BELOW Clusters of infant stars formed in a ring around the core of the barred-spiral galaxy NGC 4314; a close-up reveals dust lanes, smaller bars of stars, and an extra pair of spiral arms packed with young stars.

the huge Mount Wilson telescope, he identified whole groups of galaxies, and found that each one was receding from another. This proved that the universe was vaster than all previous estimates, and is continually expanding. Hubble discovered that the galaxies were moving farther apart based on the light they emitted. He found that other, more distant galaxies were receding more quickly.

This discovery excited scientists. Georges Lemaître, a Belgian astrophysicist, suggested in 1927 that if this movement were reversed, all the galaxies would come together in one spot at the center of the universe. Perhaps the universe had begun from what he termed a cosmic egg, by an enormous explosion that accelerated fragments in different directions throughout space? This became known as the big bang theory.

Later, a rival group of astronomers devised the steady state theory. This suggested that as the galaxies receded from one another, new material was created from interstellar gas to fill the gap, so the structure and appearance of the universe never changed. A third theory was the oscillating universe. It suggested that gravity would slow down the vanishing galaxies, which would plunge back into the center of the universe, to trigger a huge explosion and yet another expansion.

Current evidence favors the big bang theory. The colossal distance of some very distant sources of the radiation picked up by radio telescopes show that this radiation began its journey to Earth millions of years ago. Older radio sources are more numerous than objects closer to Earth, which means the structure of the universe has changed over time. A new discovery, made in 1965, reinforces the big bang theory. Scientists found that space was slightly warmer than the theoretical temperature of absolute zero. Astronomers think the heat energy was left from the original big bang. Furthermore, none of the receding galaxies seems to be slowing down, casting doubt on the idea of the oscillating universe.

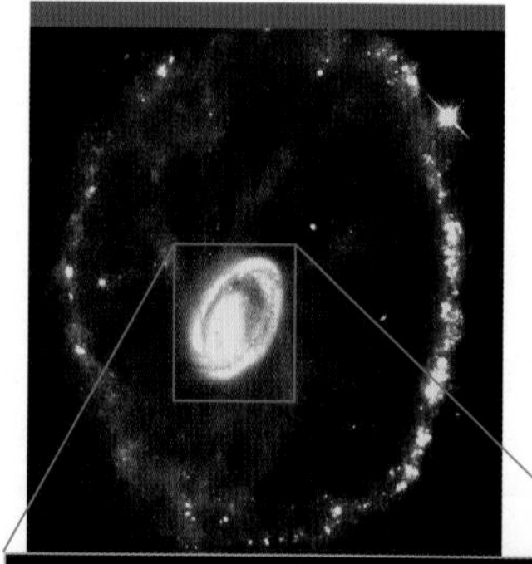

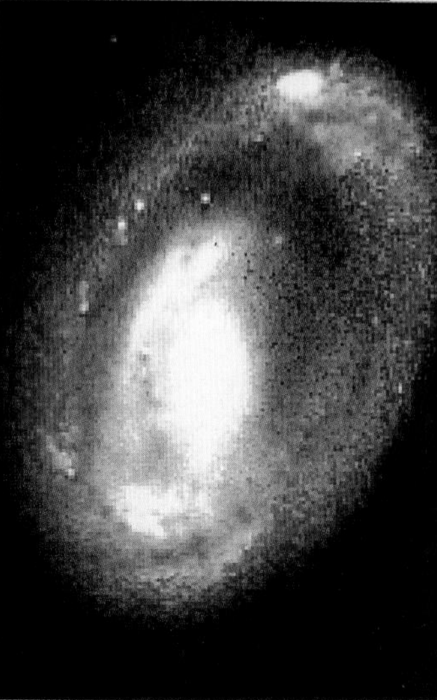

ABOVE Detail of a cartwheel galaxy including a close-up image of the nucleus with comet-like knots of gas.

LIFE CYCLE OF A STAR

Stars are born from huge clouds of dust and gas within the galaxies. As a cloud shrinks under the gravitational forces of the particles from which it is made, some areas become denser than others. These collapse into globules of matter that will form stars. When the particles collide with one another at an increasing rate, the resulting friction causes this matter to glow hotter and hotter. Eventually, after perhaps millions of years, the temperature rises to the point where a nuclear reaction begins.

Most of the atoms in the growing star are hydrogen, and the vast pressures and temperatures convert these to helium. Once this reaction stabilizes, a star can burn on for thousands of millions of years. Larger stars burn faster and more fiercely than smaller ones.

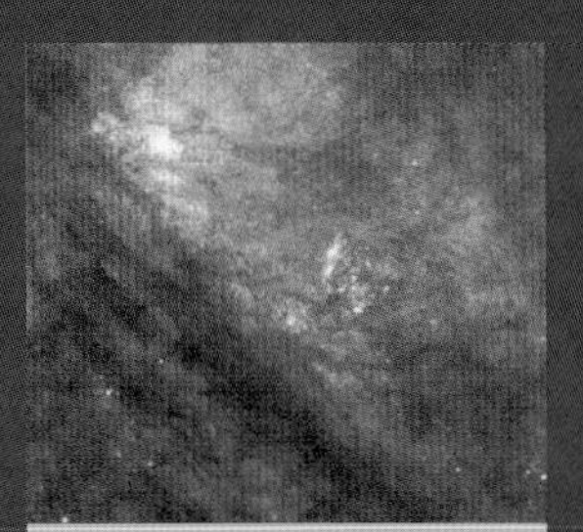

ABOVE Behind a dusty veil lies a cradle of star birth.

BELOW Image of the planetary nebula NGC 7027 showing new details of the process by which a star like the Sun dies.

The life of larger stars is limited when the reserves of unburned hydrogen at the core begin to run out. The fiercest reactions move from the center to the outer layers, with its hydrogen reserves. The star grows hotter and larger, to become a red giant.

Some stars are large enough for the outer shell of hydrogen gas to continue heating the central core, which is now mostly helium. When the temperature and pressure are high enough, another nuclear reaction breaks down the helium atoms, releasing still more energy and converting the core into carbon. This reaction then spreads outward in another red giant stage as the helium is used up. The central core cools, turning the star into a white dwarf. Finally the star dies down into a black dwarf of intensely dense ash.

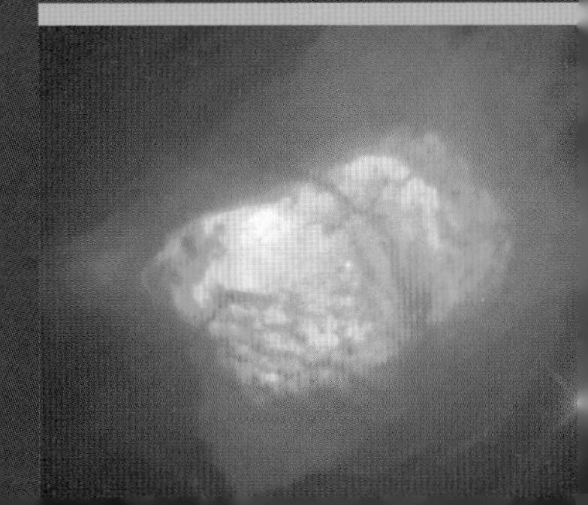

INTERPLANETARY DISTANCES

The colossal distances of interstellar space become almost meaningless when expressed in millions of millions of miles or kilometers, so astronomers needed a much larger unit of measurement. Since the speed of light is some 186,000 miles (300,000 kilometers) per second, then light travels a distance of approximately 5,874,989 million miles (9,460,530 million kilometers) in a normal year. The light year became the almost universal measurement of distances in space, partly because it helps us understand how long light from distant objects will take to reach Earth. In other words, light from a galaxy two million light years distant would take two million years to reach an observer on Earth, so we see that galaxy as it was two million years ago. To see it as it is now, an observer would have to wait another two million years!

More recently, distances have usually been expressed in even larger units called parsecs. Each parsec is a distance so vast that seen from one parsec away the width of Earth's orbit around the Sun (186,000 miles or 300 million kilometers) would appear as an angle of 1/360 of a degree. Each parsec is equal to 3.26 light years, or 19,170,000 million miles (30,856,000 million kilometers).

BELOW Images of the six planets in our solar system put together in a composite scene.

SUPERNOVAS, NEUTRON STARS AND PULSARS

The largest stars do not die quietly. Their higher gravity increases the temperature of the central core to awesome levels, so carbon is consumed to produce a series of different elements, and finally iron atoms, at its core. By this time, the internal reactions are consuming more energy than they produce. The star collapses and then explodes as a supernova, increasing its brightness up to a billion times, and hurling more than half its matter out into space. The most recent known supernova was seen on February 23, 1987, in the Large Magellanic Cloud, a neighboring galaxy to the Milky Way.

The dense central core is left behind, shrinking from the blast of the explosion and its own internal gravity. These fierce compressive forces crush the atoms into neutrons, and a star once larger than the

CONSTELLATIONS AND GALAXIES

Once astronomers discovered that the Milky Way was a spiral galaxy, with a series of arms radiating from the center, they could fix the positions of some of our nearest neighbors. For example, stars within the constellations of Orion and Cygnus are located on the same arm of the Galaxy as the Sun. Stars in the constellation of Perseus are located on a different arm, and those of Sagittarius on another arm.

The entire galaxy is spinning about its center, with each revolution taking about 225 million years. The oldest stars, found at the center of the galaxy, are some 13,000 million years old. As the galaxy's rotation accelerated over time, its motion caused it to flatten out into a spiral disc. Each arm contained huge amounts of dust and interstellar gas. As these shrank under gravity, they eventually united to form younger stars, like the Sun.

Galaxies vary widely in size and the number of stars. Spiral galaxies, such as the Milky Way, vary between 1,000 million stars and a thousand times this number, and range in breadth from around 20,000 light years to several hundred thousand light years. Elliptical galaxies can measure several million light years across, and contain as many as 10 million million stars. Distances vary from the two million light years separating Earth from the Andromeda galaxy to the 8,000 million light years of the farthest visible galaxy, code named 3C123.

THE RISE OF THE ROCKET

The development of the vehicle that made space travel possible for the first time.

The idea of space travel has fascinated people since they first studied the heavens. As early as the second century AD, writers began crafting plots in which people visited the Moon, Mars or other planets. More recently, plots about space travel and aliens have inspired filmmakers, too. But years of hard work were needed to turn space travel into a reality. Only one vehicle could deliver the power and speed necessary to escape the pull of gravity. A rocket could produce enough thrust, outside the atmosphere, to carry a payload into a stable orbit around Earth, or to journey toward other planets.

BELOW A replica of the Congreve bombarding frame of 1806.

Rocketry actually began with firecrackers and rocket-powered arrows, which the Chinese invented during the 13th and 14th centuries. This knowledge spread to the Middle East and throughout Europe. Rockets were used as war weapons. Sir William Congreve developed military rockets for the British Army in the Napoleonic Wars. His rockets weighed up to 60 pounds (27 kilograms) of either high explosive or flammable material, and were launched from folding bombarding frames like ladders.

ABOVE Konstantin Tsiolkovski, father of modern rocketry (1857–1935) in his study in 1919.

Military rockets became increasingly important during the 1800s, and Britain used them against the United States during the War of 1812. In September 1814, their use of rockets in an attack on Fort McHenry guarding the entry to Baltimore harbor was immortalized in the line "The rockets' red glare" in "The Star-Spangled Banner."

Unfortunately for the prospects of space travel, all these military rockets burned solid fuel. Solid fuel was easy to handle and safe to store for long periods. But it was heavy for the amount of thrust it produced, and also impossible to control. Once the fuel ignited, it burned until it was used up. There was no way to reduce or interrupt the thrust to steer a rocket-propelled craft, or to land it safely at the end of each flight.

Turning a military weapon into a spacecraft was a long and tedious process. It began with Konstantin Tsiolkovski, a former schoolmaster and spare-time researcher. While trying to design an all-metal airship, he built the first wind tunnel in Russia in order to study aerodynamics and streamlining. Tsiolkovski was the first to see how a rocket would operate beyond the atmosphere. As a burning rocket ejects hot gas backward, it obeys Newton's third law of motion, in producing an equal and opposite reaction to drive it forward. Unlike a jet engine, it does not need an atmosphere that the gases can impact against to create thrust. Therefore this device will work in a vacuum.

As early as 1903, Tsiolkovski proposed that space travel would require liquid-fueled multistage rockets.

ABOVE A cover from an early version of *The War of the Worlds*, the work that inspired Goddard to create rockets.

Liquid fuel was vital for the controlling of the thrust. The weight could be reduced by dividing the rocket into stages that were dropped off in flight, as the fuel contained by each stage was used up. Amazingly, Tsiolovski developed these visionary ideas while the automobile was still unreliable and the Wright Brothers were still working on their first fragile flying machine.

When he died in 1935 at age 78, his ideas were still untried. By then, Robert Hutchings Goddard, an American physicist, was also working on liquid-fuel rockets. His interest in rocket-propelled space travel had been sparked by reading H. G. Wells' science fiction classic, *The War of the Worlds.* While growing up in Massachusetts, Goddard wrote in his diary that he "imagined how wonderful it would be to make some device which even had the possibility of ascending to Mars." This dream became his life's work.

Goddard proved Tsiolkovski's prediction, that a rocket could deliver power in a vacuum. He determined the power output for various fuels, including both liquid oxygen and liquid nitrogen. By the 1920s he was using a rocket motor burning a mixture of gasoline and liquid oxygen. During a static laboratory test Goddard performed in 1925, his liquid-fueled rocket succeeded in lifting its own weight. Then in March 1926, it flew for the first time, from a field in Auburn, Massachusetts.

From 1930, Goddard's work attracted enough support to run a workshop and test facility in Roswell, New Mexico, where weather conditions were better. By 1935, his liquid-fueled rockets were exceeding the speed of sound and soaring more than a mile (1.6 kilometers) above Earth. He had already produced designs and patents for fuel pumps, cooling systems and multistage rockets, when World War II broke out. The US Government had no official interest in rocket development, so Goddard worked instead on a booster for accelerating seaplanes to takeoff speed.

When Goddard died of cancer four days before the Japanese surrender in 1945, his life's ambition was still unfulfilled. Ironically, during his final years, German rocket pioneers had used his ideas to develop new weapons — and with devastating results. Like Goddard, Hermann Oberth had read science fiction as a young man, in his case, books by Jules Verne. While serving in the Austrian army in World War I, he designed a long-range, liquid-fueled rocket. The war ministry rejected it as impractical, but Oberth continued his work. In 1923, he published a book showing how rockets could propel a spacecraft and he made suggestions for electric propulsion. His first liquid-fueled

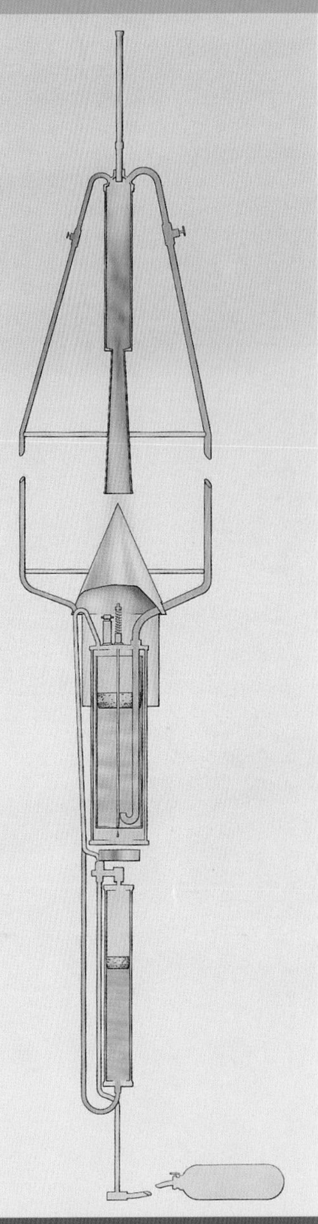

ABOVE A sectional diagram of Goddard's rocket.

LEFT Goddard standing beside his first liquid-fueled rocket on May 16, 1926.

ABOVE Wernher von Braun, who worked on the V2 rocket.

rocket was launched near Berlin, Germany, on May 7, 1931.

In 1938, Austria was annexed to Nazi Germany. Oberth himself became a German citizen in 1940. The next year, he joined a scientific team, led by Wernhner von Braun, working at Peenemünde, on the Baltic coast of northeastern Germany, to aid in the development of ballistic rockets.

As this program began, the first static tests failed. The fuel mixture of alcohol and liquid oxygen was so volatile that the slightest shock or spark would cause a violent explosion. Even when the rocket worked properly, it was very hard to control. A gyroscope was fitted in the nose and movable control surfaces to the fins, but these had little or no effect when the rocket was accelerating at the start of a flight.

By early 1935, the *A2* was ready for testing. The first two rockets, launched from a site on an island on the Baltic coast in Germany, soared over 6,000 feet (1,830 meters) high. Impressed, government officials gave more attention and funding to Peenemünde.

The *A3* was a far more ambitious design than the *A1* and *A2.* Like the *A1,* its engine burned a mixture of liquid oxygen and alcohol, but it developed no less than 3,000 pounds (1,360 kilograms) of thrust. Weighing three-quarters of a ton and standing higher than 20 feet (6 meters), this was the largest rocket yet designed.

However, the *A3* never made a successful flight. There was no way to steer this bigger and heavier rocket during the critical first few seconds of flight, when it was lifting off. While the rocket was supported on its own fiery exhaust, the slightest instability would cause it to tip over before it could gain sufficient speed.

By then von Braun's team was creating the even more ambitious *A4,* which weighed more than 12 tons and stood higher than 46 feet (14 meters). This

ROCKET-PROPELLED TERROR

By autumn 1944, World War II was almost over, and Germany was being crushed from both east and west. Only the unstoppable V2 offered a secure way to fight back. On September 6 the first armed V2 struck Paris. Two nights later, the first V2 aimed at London exploded in suburban Chiswick; 16 seconds later another hit Epping — the beginning of six months of terror and destruction.

The completely unpredictable arrival of a V2 was utterly terrifying and there was no effective defense against this new weapon. Each warhead killed an average of five people. Moreover, the missile flew faster than sound, so the noise of its approach could only be heard after the explosion. The Germans also used movable launchers, which proved virtually impossible to spot.

BELOW V2 damage in Antwerp, Belgium in 1944.

By early April 1945, when German troops had finally been pushed back far enough to put Allied cities out of rocket range, thousands of V2s had been aimed at Paris, London, and Antwerp. Of the 1,359 V2s they fired at London, the Germans were aware of only 169 that failed en route. British records showed that 1,115 exploded on British soil, but accuracy remained a problem for the rocket. Although London was a huge target, less than half the missiles, 501 in all, actually fell within the London Civil Defence Region.

was to carry a one-ton payload more than 150 miles (290 kilometers). To achieve that feat, the motor, burning the same volatile mixture of liquid oxygen and alcohol, would have to develop 60,000 pounds (27,200 kilograms) of thrust. The control problem was now even more urgent.

To solve it, von Braun's team built a smaller scale model, the *A5*. It was similar in size to the failed *A3* but the framework used the same techniques as the light, but strong, frames of the giant Zeppelin airships. The control system was revised, so the missile could be steered in the first few seconds of flight by movable vanes situated in the rocket exhaust. By 1939, a series of *A5*s had been produced and flight-tested over the Baltic. The rockets reached heights of 35,000 feet (10,670 meters).

The first complete *A4* was finally ready for firing in June 1942. By then, it had acquired a new purpose: as a weapon, aimed to unleash its cargo of almost a ton of high explosives on Allied cities. Its new name, V2, first stood for *Versuchsmuster 2*, or *Experimental Type 2*. For propaganda purposes, this was changed to *Vergeltungswaffe 2*, or *Revenge Weapon 2*.

BELOW The *A4* rocket, dubbed by the Nazi propaganda machine the *Revenge Weapon 2*, or "V2."

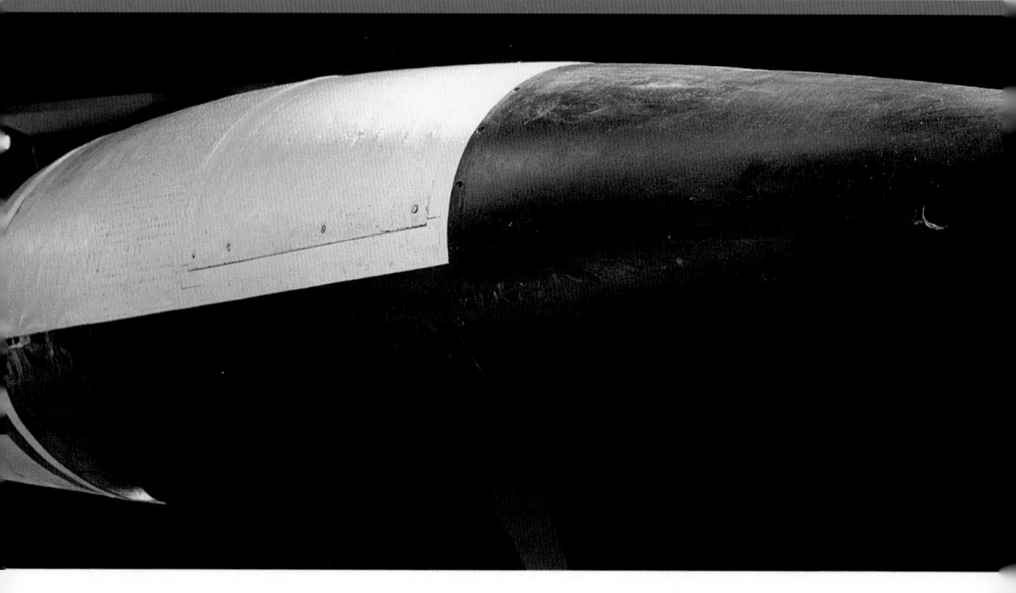

ABOVE A rocket used in the film *Destination Moon,* based on the shape of the V2.

Turning the experimental rocket into a reliable war weapon was much more difficult than renaming it. Its first flight ended when the fuel pump in the motor failed. After a few seconds of acceleration, the rocket toppled over and exploded. Then, on August 16, 1942, the second V2 was launched. Again the motor failed, though this time the missile accelerated past the speed of sound before crashing back to Earth. Only on October 3, with the launch of the third V2, did the rocket perform as designed. The missile reached an altitude of 50 miles (80 kilometers) and flew almost 120 miles (193 kilometers).

When the Germans began mass-producing the V2, it had a range of 260 miles (418 kilometers). Because it flew far faster than sound, the shattering explosion of its one ton warhead on impact came with no warning. By September 1944, the Germans had attacked London and Paris with the first of 4,000 missiles they would fire against the Allies. The V2 program was limited only by the rate at which missiles could be made and the scarcity of the fuels they required, since Germany was increasingly suffering from Allied bombing and blockade. Eventually, the retreat of the German army meant that potential launching sites were too far from their intended targets.

As the war ended, the Allies knew that German rocket development would be a trump card in postwar competition. As they scrambled for the rockets, the scientists and the documents relating to rocket research, the former Allies switched to pursuing their own interests. The British test-fired several V2s from a range in their part of Germany, but it was clear that only the Americans and the Russians had the resources and the determination to mount a major rocket-development program. The US offered the German scientists in their hands, including most of the control and guidance specialists, contracts to work in America. Those in Russian hands, including key rocket-propulsion specialists, were simply put under armed guard and taken by train to the Soviet Union, to continue their work for new masters.

So, in the first years of the uneasy peace, V2s were launched from new sites on both sides of the world. The pioneering work at Peenemünde would help the Russians to place the first satellite in orbit around Earth, and put the first man into space. On the American side, it would provide the basis for the Saturn rocket, which would propel astronauts to land on the surface of the Moon. Much work still lay ahead for both sides, however.

The Americans had seized 100 assembled V2s, ignoring British protests that wartime agreements entitled them to half of the missiles. They moved the missiles and their team of German scientists to the White Sands Proving Ground in New Mexico. The first American V2 passed a static test on March 14, 1946. A second rocket was fitted with a payload of

BELOW The V2 in the United States, at the Air Force Missile Test Center.

instruments, and launched to a height of 67 miles (108 kilometers) on June 28, 1946. A long series of test launches went well. Then, on May 29, 1947, a V2 veered across the border into Mexican airspace and crashed into a cemetery, fortunately missing the crowds at a local fiesta. After this mishap, the program continued successfully.

Meanwhile, the first Russian V2, launched on October 30, 1947, flew for 185 miles (298 kilometers), then crashed back to earth right on target. The missile race and the space race had begun. Both started favorably for the US, as it launched a V2-boosted WAC Corporal rocket on February 28, 1949 to a record altitude of 259 miles (417 kilometers). That achievement remained unmatched for eight more years. Racing ahead, the Soviets held the lead during the next decade before the US team and their German colleagues achieved the ultimate triumph: the development of the multistage Saturn liquid-fueled rocket, which would carry the Apollo moonflights.

UNRAVELLING THE V2 SECRETS

Though the British could not prevent the V2s from reaching London, they did know about the threat. They had inspected parts of a crashed V2 that veered off course from Peenemünde. That rocket came down in Sweden on June 13, 1944, less than three months before the V2 attacks began.

More valuable information came from the Polish Resistance, who had been watching the Germans test V2s from a range at Blizna in eastern Poland. Time after time they would rush to a crash site, only to find German patrols had removed vital components for analysis. Finally on May 20, 1944, the Poles arrived first after a V2 crashed near Sarnaki, a village on the banks of the river Bug. Knowing they had to move quickly before the Germans arrived, the Resistance men rolled the wreckage into the river and screened it with a herd of cows to stir up the mud.

The Germans did not find the rocket, so the Poles were able to dismantle it in peace, and make detailed drawings. The Allies agreed to a daring plan. They would send an unarmed DC3 transport plane to make a night landing at an unused airfield deep in German-occupied Poland, and pick up the parts and the drawings.

Carrying that precious cargo in a sack slung over his shoulder, a Polish Resistance agent cycled 200 miles (320 kilometers) through the retreating German army. He and the sack reached the landing ground on July 26 and awaited the DC3 that had left from Brindisi, Italy. Then, to the horror of the Poles hidden alongside the airfield, two German fighters arrived. Had they been betrayed, and were these Nazi fighters waiting for the RAF plane?

ABOVE The first launch from Cape Canaveral of a V2 rocket.

To their intense relief, the fighters started up and took off as darkness fell. Just before midnight the DC3 landed, and the V2 parts and drawings were hastily loaded on board. But when the aircraft tried to take off, its tires stuck in the mud from recent heavy rain. As the DC3's engines ran at full power, their roar echoed across the silent countryside. The courageous plan seemed doomed.

Local farmers saved the day. They tore down nearby fences and laid the planks on the field to take the weight of the aircraft. With spades and their bare hands, they scrabbled away at the mud around the wheels. Finally the plane began to move, slowly reaching takeoff speed. As it vanished into the night, the German forces arrived, running into a resistance ambush. Within three days the parts and the drawings were safely in London.

3

LAUNCHING THE FIRST SATELLITES

The international space race took off as the Soviet *Sputnik 1* rocketed into orbit, the first satellite in space.

ABOVE AND OPPOSITE
The world's first artificial Earth satellite, launched by the Soviet Union on October 4, 1957.

October 4, 1957, will always be a historic day in the Space Race. On that evening, the Soviets managed to put the first artificial satellite into orbit around Earth. Named *Sputnik 1* (Russian for "fellow traveler"), the satellite was spherical in shape and weighed 184 pounds (83 kilograms). It was placed in an elliptical orbit so that its height above Earth's surface varied from 584 miles (940 kilometers) to 143 miles (230 kilometers). The regular bleeping of its battery-powered radio transmitter could be heard all over the world — a constant reminder that Russia now led the space race.

The Russians had clearly signalled their plans. Their postwar rocket program focused on building more powerful versions of the V2, with engines that would eventually be able to threaten the US. By 1957, they produced a rocket that could carry a two-ton warhead more than 4,000 miles (6,930 kilometers).

Now the Soviets needed a way to reveal their dramatic new weapon. Since 1957 had been designated as International Geophysical Year, Soviet leaders decided to use their new rocket to launch a satellite. This would prove that Russian technology reigned supreme and also would show the rocket's force, giving them a big political advantage, especially over the US.

When Russia announced in June 1957 that a satellite was ready, the Americans were skeptical. Their own progress had been much slower, as they

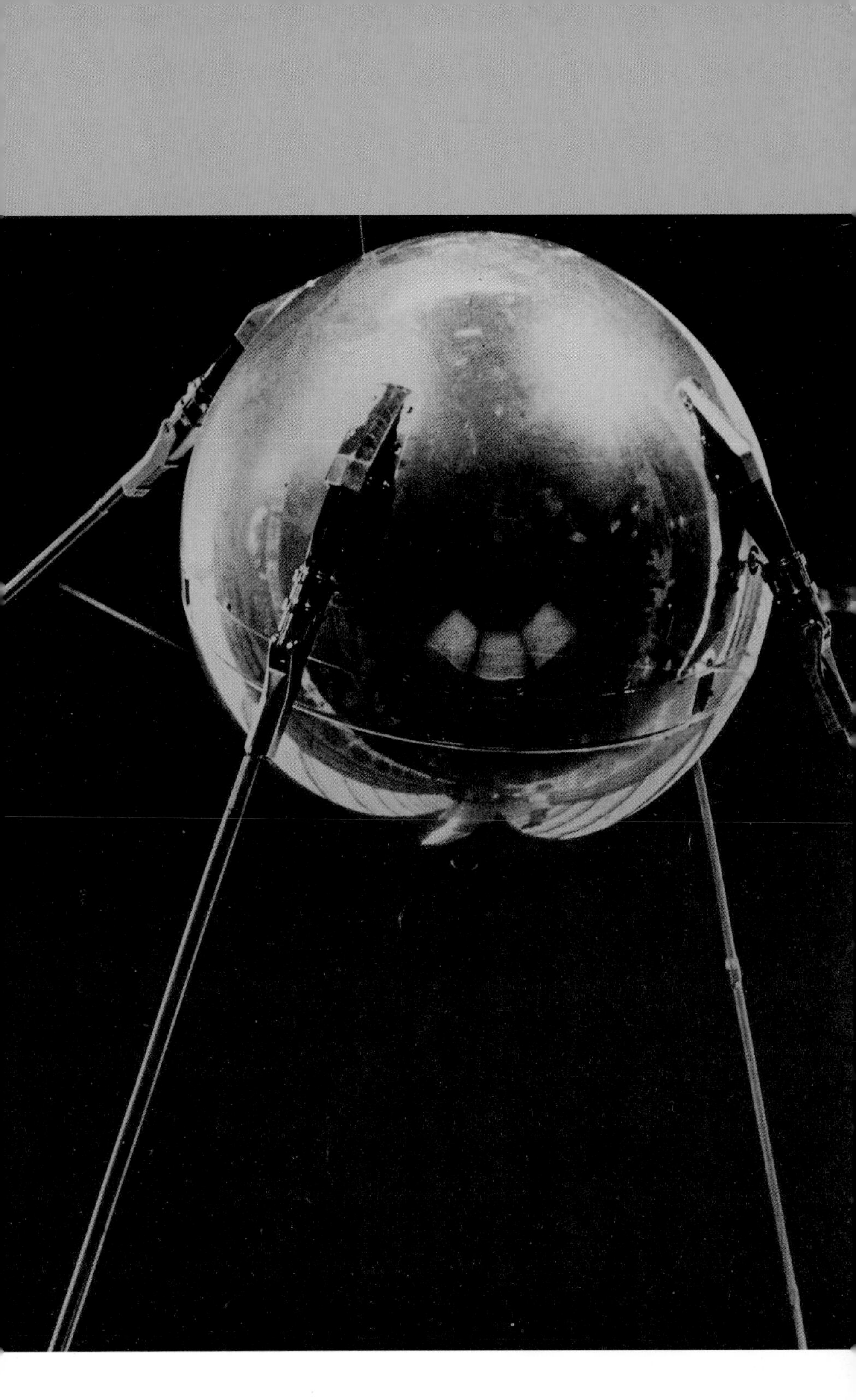

used their original stock of German V2s to design new missiles that might one day be able to deliver warheads to the opposite hemisphere. In the meantime, they hoped that American nuclear weapons, carried aboard manned bombers, would deter Russia from launching an attack.

After *Sputnik 1* began its historic orbit, people learned that the rocket consisted of two stages, each one fitted with four rocket engines. It had achieved a speed of 18,000 miles per hour (28,900 kilometers per hour). *Sputnik 1* stayed in orbit for three months before Earth's gravity overcame it. The satellite burned upon re-entering the atmosphere.

In 1955, the US had announced plans to launch Earth satellites in two years as part of the International Geophysical Year program. Now the whole world could see they had been outdone. Ironically, their problems were mainly self-imposed. The German experts under Wernher von Braun were working for the US Army on a promising rocket called the Redstone, while the US Air Force had the Atlas, which had been developed as a potential ballistic missile. The US Navy had developed a three-stage satellite-launching rocket called Vanguard, which was based on a civilian research rocket called the Viking.

Amazingly, the administration chose the Vanguard, the least promising of the three projects. Its payload limitations were so severe that the first US satellite weighed just three and a half pounds (1.6 kilograms), less than 1/50th of the weight of *Sputnik 1*.

Even before that small step was taken, the Russians increased their lead on November 3, 1957, by launching *Sputnik 2*, a satellite six times the size of *Sputnik 1*. On board was a small dog named Laika, the first living creature to voyage into space, along

LEFT Failure of the US Vanguard launch vehicle on its launch pad on December 6, 1957.

ORBITS AND FOOTPRINTS

In early satellite launches, scientists looked for every possible way to assist the satellite into orbit. For example, Earth's spinning on its axis meant that launching sites near the equator were moving rapidly from west to east relative to the surrounding space. If a rocket was launched vertically, it would be traveling eastward as Earth rotated beneath it. If the rocket was then steered to an eastward trajectory, it and the satellite it carried would reach a greater velocity relative to Earth than if it had been launched to the west. That meant it could be established in a higher orbit. This was called "hitching a ride," and was used in most of the early satellite launches.

Later, launching equipment became more reliable, allowing the orbits to be selected more accurately, so there were more ways to cover Earth's surface. Satellites designed to relay communications between different ground transmitters, like TV satellites and military communications satellites, are launched on an orbit that parallels the equator. Scientists choose the height of the orbit so that the satellite's speed through space is matched by the rotation of Earth.

This means the satellite effectively hovers over the same spot on the surface in what is called a geosynchronous orbit. The intention is that the satellite's footprint, or the area covered by its reradiated transmissions, will remain the same in spite of the rotation of Earth.

ABOVE The European rocket *Ariane 5* on its launch pad.

Satellites that monitor weather systems or collect information on natural resources or military deployments need to cover the whole surface in successive orbits. These are launched in an orbit that takes them over the North and South Poles on each circuit. As the planet rotates beneath their path, they cover its entire surface on successive passes.

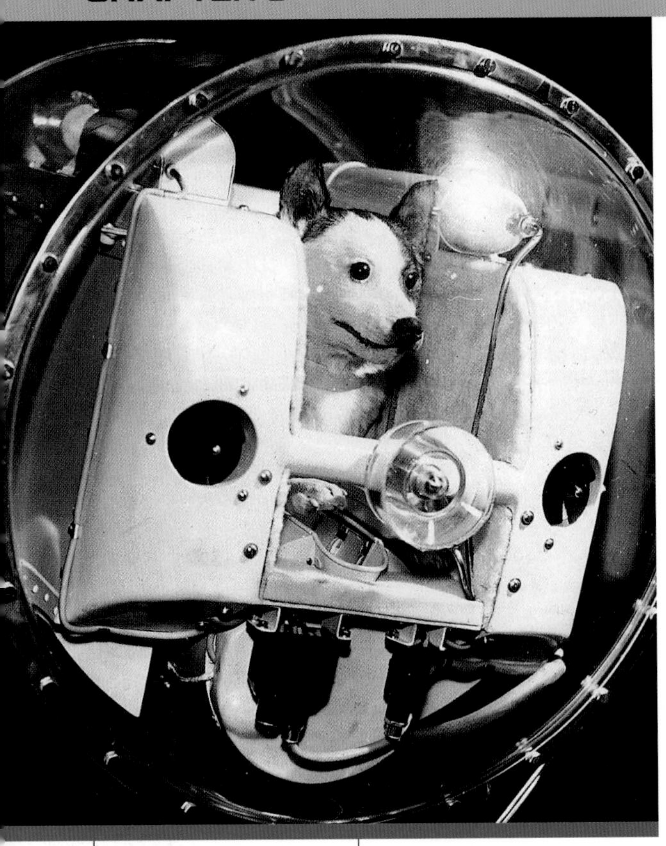

with a life-support system and instruments to record the dog's heartbeat, temperature and other vital functions to help show whether humans could survive in space. A transmitter relayed the data back to Mission Control.

ABOVE A model of the dog Laika in the container that was sent up with *Sputnik 2* on November 3, 1957.

OPPOSITE A jubilant report in an Alabama newspaper about the successful launch of the *Jupiter-C* missile on January 31, 1958.

The first attempt to launch Vanguard and its tiny satellite, on December 6, was a disaster. The rocket rose off its launching pad, then toppled over and exploded. Yet hope lay on the horizon. Von Braun and his team had been developing *Jupiter-C*, a multistage version of the Redstone. It weighed three times as much as the similarly-sized Vanguard and its extra booster stage contained a battery of 11 solid-fuel rocket motors. The first of these rockets had been launched in 1956, reaching a height of 600 miles (965 kilometers) and a range of 3,000 miles (4,830 kilometers), which was far better than any other American rocket. Quickly US officials switched responsibility for launching America's second satellite, the 31-pound (14 kilogram) *Explorer 1*, to the army and *Jupiter-C*.

Von Braun and his team triumphed on January 31, 1958, when *Explorer* blasted off without a hitch and the satellite was safely placed into Earth orbit. Though it was still much smaller than the Soviet satellites, America had a strong lead in electronics and instrumentation. Almost immediately once it was in orbit, *Explorer's* instruments detected so many electrically charged atoms that the recorders went off the scale. The satellite had encountered what became called the van Allen radiation belts, after James van Allen, the scientist who directed the experiment. These were belts of protons and electrons from the Sun, carried on the solar wind and trapped by Earth's magnetic field.

By the time *Explorer* was launched, the Soviets were ready for their next big step. On May 15, 1958, *Sputnik 3* rose into orbit, exploiting the sheer power of the Russian rockets. Fully two-thirds of the satellite's ton and a half weight was taken up by instrumentation and transmitting facilities. Clearly, the Russians would soon send the first human astronaut into space.

The Americans realized that their Voyager program would never have enough power to do this. They looked toward von Braun's *Jupiter* rocket and a reusable hybrid rocket plane called the Dyna Soar, being developed for the air force. Although this would never be carried through to a prototype, the basic principles would eventually be developed into the space shuttle (as described in chapter six).

In order to limit the influence and expense of big defense-based contracts, the US administration sought to keep its basically civilian space program separate from military rockets. President Eisenhower and his advisors chose the National Advisory Council for Aeronautics, which had been established in 1915, to coordinate aeronautical research and development. In 1958, the council was renamed the National Aeronautics and Space Agency (NASA). Later, the last word was changed to "Administration" to reflect NASA's wider responsibilities.

NASA set two big goals. First, they would put an American astronaut into space as soon as possible rather than

SATELLITE EXTRA — The Huntsville Times — SATELLITE EXTRA

Jupiter-C Puts Up Moon

Eisenhower Officially Announces Huntsville Satellite Circles Globe

Wail Of Sirens Brings In Era On Space Here

Thousands Gather On The Square For Noisy Success Demonstration

9 Labs Here Aided Project Of Launching

It Took Every One To Successfully Put Up 'Moon' Vehicle

Weather Change Sped Launching

Army Reveals Second Moon Is Scheduled

Here Are The Basic Facts

Launching Hits ABMA Birth Eve

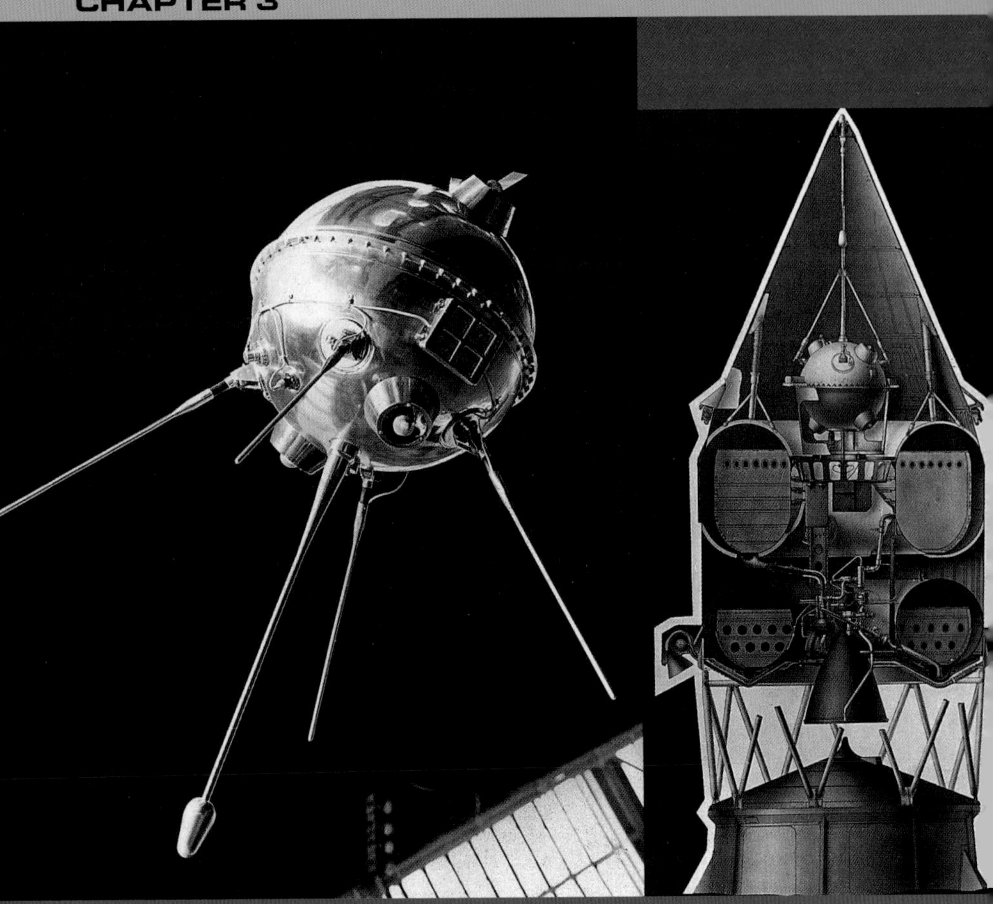

ABOVE LEFT ***Luna 2*** **automatic station.**

ABOVE A diagram of the ***Luna 1*** **space station.**

leave space exploration only to the Russians. NASA also made plans for a series of unmanned vehicles to explore the Moon. Now Von Braun and his team were working directly for NASA. They relocated at the misleadingly named Jet Propulsion Laboratory at Pasadena, California.

Meanwhile, the Russians continued to forge ahead, both in the new area of space exploration and in military uses for rockets. On January 2, 1959, the Russians launched *Luna 1*, which became the first object to escape Earth's gravitational field once and for all. *Luna 1* passed beyond the Moon's weaker gravitational influence, transmitting information all the time. Then it headed out into a permanent orbit around the Sun, to become the first artificial solar satellite. Another coup soon followed when *Luna 2*

New types of satellites
AND SATELLITE WARFARE

Satellites are also vital aids to navigation. Their elevated view of Earth's surface has enabled people to make more accurate maps. Satellite-based navigation systems enable ships, aircraft and even individuals to accurately determine their position when visibility is very poor. The Landsat series of satellites has used different types of spectral imaging to monitor agriculture and land use, forestry and mineral deposits and water supplies.

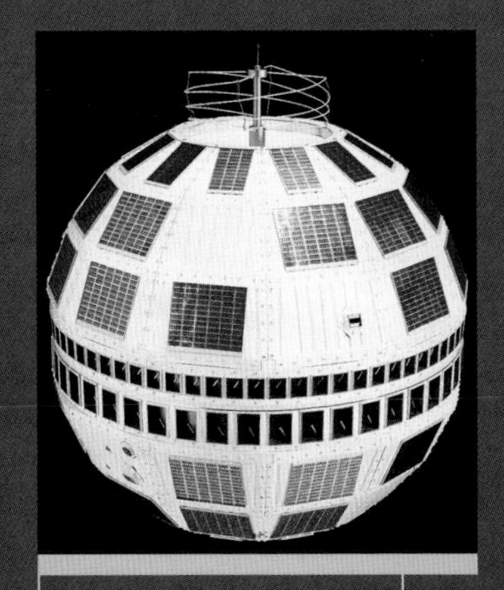

ABOVE The *Telstar 1* satellite, 1962.

The US Navy's Geosat satellite can map ocean currents and the contours of the seabed. The Laser Geodynamic Satellite (LAGEOS) was launched as long ago as 1976 into a very high orbit some 3,600 miles (5,790 kilometers) above Earth, which should give it an operating life of several million years! LAGEOS contains 426 reflectors that enable ground-based lasers to be used to measure tides, continental drifts and even slippages along geological fault lines, all of which make earthquake prediction more reliable. Finally, communications satellites now relay telephone conversations and color television pictures between continents and over wide areas of Earth.

Special satellites were developed to give the earliest possible warning of a possible ballistic missile attack. These became prime targets for any nation considering a first strike. The Russians found ways to destroy satellites by moving other controllable satellites close enough to deliver a killer punch. These would either explode at close range or launch a small cloud of metal pellets or needles into the path of the other satellite, which could destroy it, or at least make it useless. But these methods were massive and cumbersome, which limited them to low orbits. By the time they were maneuvered into an attack position, satellites in higher orbits were likely to be out of range.

For their part, the Americans also devised a series of missiles that could be launched against satellite targets. The first ones were launched from ground sites. Then, during the 1980s, they developed a Miniature Homing Vehicle (MHV), a smaller missile that could be launched from a specially equipped F-15 fighter. It was accurate enough to directly hit the target and destroy it without any explosives. As the F-15 was carried aboard US Navy aircraft carriers, the weapon could theoretically be launched from most parts of the globe. But although the fighter could launch the missile at heights of up to 10 miles (16 kilometers), the missile could still only hit satellites in relatively low orbits.

BELOW Assembly of the ***Intelsat VI*** satellite.

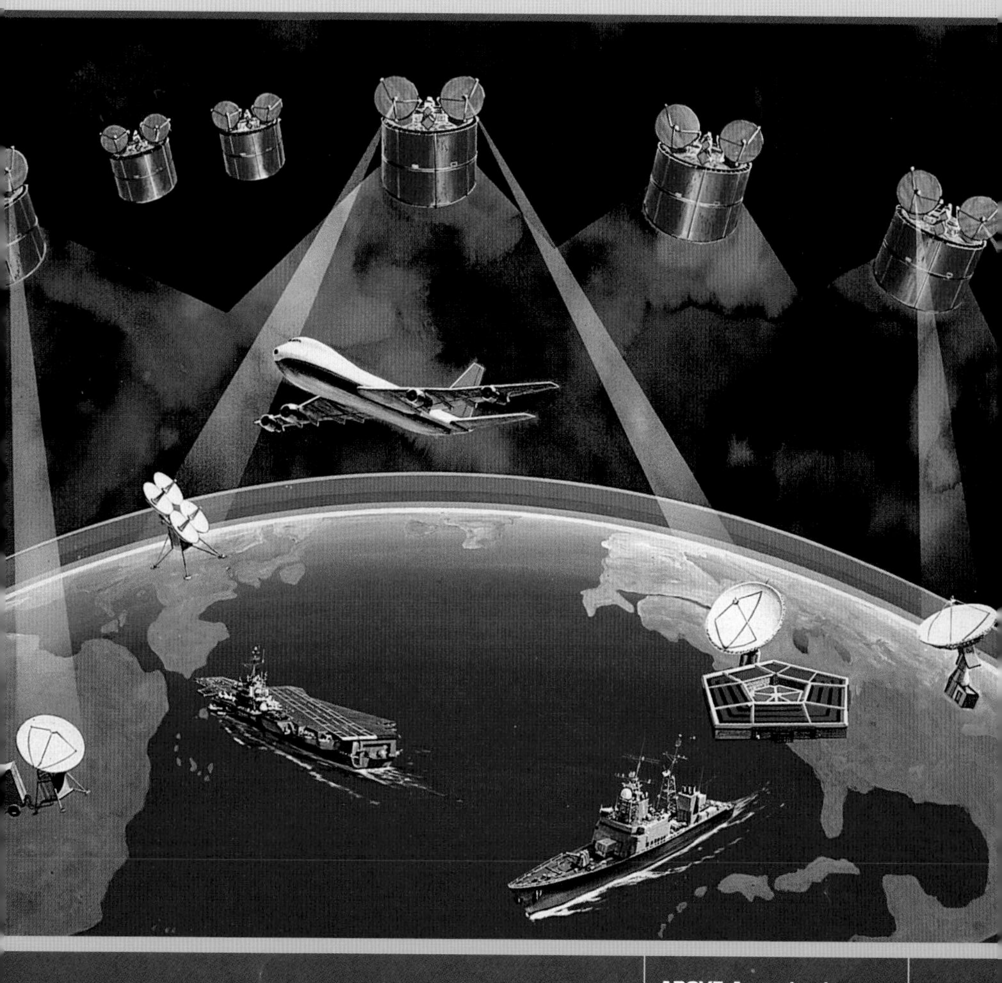

ABOVE An artists's impression of the constellations of *Defense Satellite Communications System 2* in geostationary orbit in the 1970s.

The theme was extended in the Strategic Defense Initiative, popularly known as Star Wars. This was based on a series of antimissile weapons which were themselves satellites in orbit around Earth. They included infrared telescopes to detect missile launches, and laser radars to track their flight, reinforced by high-energy spaceborne lasers to destroy the missiles by burning holes through their outer skin. It was never proved whether all this sophisticated technology would actually work during a genuine attack.

was launched in September, and aimed squarely at the Moon. It ended its flight by crashing into the lunar surface a day later. A proud moment came three weeks later when *Luna 3* was launched. Its orbit took it around the far side of the Moon, which had never been visible from Earth. *Luna 3* sent back pictures showing the first views of this region.

By comparison, American efforts looked dull. Nevertheless, they were developing more expertise and equipment for observation and research. In 1960, the US launched two new, special-purpose satellites. *Echo 1* was a plastic balloon coated with aluminum that could reflect radio signals beamed at it from the Earth's surface and relay them back to Earth at a spot normally inaccessible because of the curvature of Earth's surface. Next came TIROS, which was launched in April 1960 to become the first weather-monitoring satellite. It carried television cameras powered by batteries charged by energy from solar cells. These cameras were used to relay data on weather systems and watch for the development and growth of storm systems.

By 1961, America had progressed so quickly that they very nearly beat the Russians in putting the first human into space (see chapter four). Furthermore, after John F. Kennedy was elected President in 1960, his administration radically changed NASA's objectives. Instead of exploring the Moon with remote, unmanned probes, NASA resolved to land astronauts on the moon before the end of the 1960s (see chapter five).

During that decade, the Americans modified several ballistic missiles to provide more powerful launch vehicles. These included the *Titan II*, which had two solid-fuel booster rockets strapped to the liquid-fueled missile, the liquid-fueled Thor

LEFT The launch of Britain's first satellite, *Ariel 1* in April 1962.

intermediate-range ballistic missile, and, later, both Atlas and Titan missiles.

Today's rocket launchers are so reliable that they have an outstanding safety record. Now scientists focus on the orbiters themselves, to develop more sophisticated instruments for gathering information, and to lower the cost of satellite launching. Other nations besides America and Russia also have placed satellites in orbit, either by developing their own launch vehicles or by cooperating with countries that have this technology.

France successfully launched a satellite in 1965, followed by Japan and the People's Republic of China in 1970, and the United Kingdom in 1971. More recently, the Western European countries set up the European Space Agency program to develop the Ariane booster rocket. This rocket uses solid fuel in its first two stages, and a cryogenic engine in the third and final stage. It is powerful enough to launch two satellites at once, each weighing one and three-quarter tons, or one satellite weighing up to four and a half tons.

Two projects for future low-cost launchers are currently being developed in the United States. The Lockheed Martin Corporation is working on the Reusable Launch Vehicle concept, which operates on a similar principle to the space shuttle, taking off vertically on rocket motors and landing horizontally like an aircraft. In this case, though, it has no crew aboard, and has no separate boosters to be jettisoned during flight. Meanwhile, Boeing heads a multinational consortium that is converting an offshore oil-drilling platform into a mobile launch pad that could be towed to the equator, where the satellites would receive the maximum benefit from Earth's own rotation as they head into space.

LEFT The first of a series of Ariane launches at the French Guiana facility on December 24, 1979.

4
HUMANS IN SPACE

Rapid developments in technology led to longer and more ambitious flights, space walks and the possibility that humans might land on the moon.

What nation would be the first to put a human astronaut into space? In 1957, the Russians had taken the lead in the Space Race with the spectacular launch of *Sputnik 1.* Then the Americans began to develop superior electronics, instrumentation and control. The Soviets triumphed again by launching animals into space. After Laika became the original space dog, *Sputnik 5* launched and recovered two dogs on August 20, 1960. Both dogs survived with no apparent damage.

The Americans hoped to close the gap with Project Mercury, targeted with placing a man aboard a satellite. On January 31, 1961, a spacecraft containing Ham, a chimpanzee, was lobbed into space, then brought back into the atmosphere before it completed a full orbit. The technology seemed sound, and astronaut Alan Shephard was training for a similar suborbital flight just six weeks later.

But one problem sounded an alarm: The Redstone booster that had propelled Ham into space had provided slightly too much thrust so that the spacecraft was delayed in returning to Earth. Though Ham was unhurt, officials decided to postpone Shephard's flight for almost two months while they performed further tests.

That decision was sensible but it damaged American prestige. On April 12, 1961, a Soviet rocket

ABOVE Yuri Gagarin, the first man in space, at the command desk of the Vostok rocket, 1961.

OPPOSITE Launch of the *Friendship* capsule in 1962.

launched a Vostok spacecraft into orbit carrying Lieutenant Yuri Gagarin. He completed a full orbit, then his spacecraft made a flawless reentry into Earth's atmosphere and Gagarin parachuted down to a soft landing on Russian soil. Once again, the Soviet Union had scored a dramatic "first" in a new round of the Space Race.

Less than a month later, on May 5, Shephard finally took off. Unlike the more powerful Soviet rocket, the US rocket could not lift a large enough payload to let the Mercury capsule descend over land. Instead, the returning astronaut was dropped into the Atlantic, where US Navy rescue helicopters picked up the capsule. Furthermore, Shephard was in space for only 15 minutes, reaching a height of 115 miles (185 kilometers). Still, his flight showed that the Americans were catching up.

To highlight this success, a second Mercury suborbital flight was launched on July 21, 1961. This time, astronaut Virgil "Gus" Grissom was on board. Then the Russians surged forward again. Starting on August 6, 1961, cosmonaut Herman Titov spent more than 24 hours in space, making 17 orbits of Earth. Titov slept part of the time and ate specially prepared meals. Later, he complained that the weightless conditions made him feel severe nausea.

By February 20, 1962, the Americans were ready for a full orbital flight. The powerful *Atlas D* rocket, adapted from an intercontinental ballistic missile developed for the US Air Force, launched the *Friendship* capsule. Astronaut John Glenn made three orbits at a height of between 100 miles (161 kilometers) and 163 miles (262 kilometers), which was less than the Soviets were achieving.

A system malfunction showed up during Glenn's flight. Instruments on the ground at Mission Control showed that the spacecraft's heatshield had been loosened. The heatshield was designed to be burned up, absorbing the friction and heat energy during reentry into the atmosphere. If it broke off, the results would be fatal.

Quickly, the ground control team searched for a solution. Instead of the normal sequence, which would jettison the retro-rocket when it had been fired, Mission Control told Glenn to keep it in place. It was mounted in the center of the heatshield, so it provided some protection and helped to hold the assembly in place. As the team waited anxiously, Glenn fired it on cue. The rocket assembly and heatshield stayed in place, and the *Friendship* capsule splashed down safely. When the capsule was examined, the heatshield was fine — so the problem had been in the monitoring instruments!

BELOW Herman Titov and Adrian Nikolayev on *Vostok 2*, August 6, 1961.

The Russians got even with two new flights of their own. On August 11, 1962, Adrian Nikolayev went into orbit for four days, and Pavel Popovitch flew in an identical craft on a similar mission the very next day. Around the world, people wondered what new and exciting voyages these adversaries were planning.

Both the Americans and the Russians now possessed more powerful launchers. The last two Mercury missions lasted longer than ever before. On October 2, 1962, Walter Schirra spent nine hours in space, but he ran into problems like those that Titov faced on the first Russian 24-hour mission. Spending a longer period in weightless conditions meant that

ABOVE A model of the *Vostok 1* capsule, 1961.

he suffered from dizziness and circulatory problems when he returned to Earth. Finally, on May 15, 1963, Gordon Cooper spent some 34 hours in space. Though this was still brief by Russian standards, Cooper was able to carry out a number of complex controlled maneuvers, steering his spacecraft by directional rocket motors. He also took the opportunity to photograph Earth's surface from his elevated viewpoint.

The last two flights in the Vostok series of single-person launches were made in June 1963. *Vostok 5* was launched on June 14, and was quickly followed by another, on June 16. This was another "first," since it carried the world's first female space traveler, Valentina Tereshkova, a textile worker. She had not been part of the original Russian cosmonaut training program, but had been recruited specially as a public relations coup. Though she had no previous piloting experience, she was an enthusiastic free-fall parachutist.

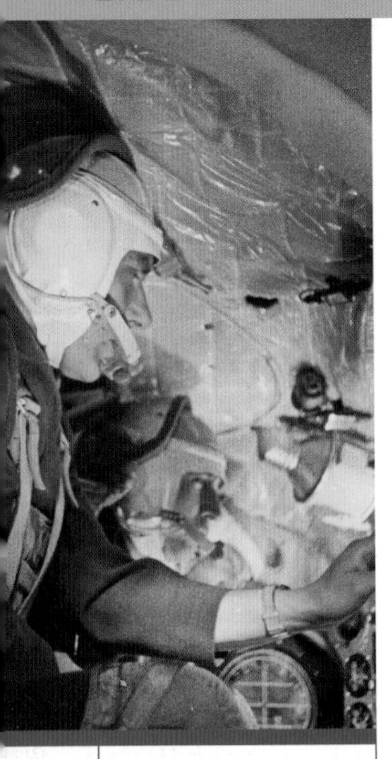

ABOVE The crew of *Voskhod 1*, October 12, 1964.

The orbits of the two flights suggested an attempt at docking, though Tereshkova's lack of piloting expertise made this unlikely. Nevertheless, at one point the two crafts were only 3 miles (4.8 kilometers) apart. During the three days, both crafts were in orbit together at times, so an even closer rendezvous would have been possible.

By this time, American plans for the Gemini two-man spacecraft were far advanced, and they hoped to excel in their orbital flights. Before the first Gemini flight, however, Russia launched a large new spacecraft on October 12, 1964, with three crewmembers. The *Voskhod 1* was a modified Vostok weighing almost three-quarters of a ton more than its predecessor. It was the first spacecraft to use an ion engine, where a jet of charged particles (ions) and electrons is accelerated in an electric field and fired out into space. This produces a similar effect to a rocket engine, from a much lighter installation.

The second Voskhod flight, on March 18, 1965, increased its endurance by carrying only two men. Some of the savings in weight and interior space were used to carry space suits, which had been left off *Voskhod 1,* and also to install an air lock to let the crew leave the craft while in orbit. On the launch day, Alexei Leonov climbed into his suit and inflated it to a low pressure of 6 pounds per square inch (0.42 kilograms per square centimeter). This was less than half the normal atmospheric pressure at sea level but would allow him to maneuver inside the suit, in the vacuum of space.

Leonov went through the airlock to emerge outside the spacecraft for the very first space walk. While tethered on a safety line, he was able to enjoy the sensations of floating free in weightless conditions, high above Earth. He then tried to climb in through the hatch, but had to partially deflate his suit to wriggle through the small opening.

The crew faced a new problem when the retro-rocket that should fire automatically to bring them out of orbit failed to work. In order to start the

GEMINI – TEST RUN FOR THE MOON

While the Russians tended to aim for spectacular "firsts," the Americans took a longer-term view after President Kennedy declared they would make a return trip to the Moon (see chapter five). The US space program switched to the painstaking process of developing and improving the techniques and technology needed for that expedition.

The Gemini project entailed frequent flights and tests, as researchers worked out the complex details and solved each problem that arose. Two manned Gemini flights scheduled for August 1965 encountered glitches that kept them on the ground. By December 4, two Geminis were finally ready to launch. Frank Borman and James Lovell took the first one into space; Walter Schirra and Tom Stafford followed 11 days later. The two spacecraft maneuvered until only feet separated them, before Schirra and Stafford took their Gemini back to Earth after just one day in space. Borman and Lovell stayed in orbit for another two days — a new space endurance record of 330 hours.

When Neil Armstrong and David Scott flew on March 16, 1966, they were able to dock with a separately launched Agena spacecraft, for the first time ever. Unfortunately, one of the maneuvering rockets onboard the Gemini refused to shut down.

RIGHT The *Gemini 4* launch in 1965.

This caused the combination of the Gemini and the Agena to tumble out of control. The two astronauts had to separate them, then shut down the offending engine. By that time, fuel was running low, so the trip was cut short, after only 11 hours in space.

Clearly, the old days of one stunning achievement after another were over. The Americans were carefully evolving an entirely new technology, solving new problems and learning the skills to make moon flights possible. On June 3, 1966, Tom Stafford and Eugene Cernan took off to practice three rendezvous with another Agena. This time the Agena's docking mechanism jammed, so they could only practice approaches to the target and undertake more ambitious space walks. After spending over two hours in space, using a rocket pack on his back to control his movements, Cernan was worn out.

BELOW **An artist's concept of a two-man Gemini spacecraft in flight, showing a cutaway view.**

ABOVE Astronaut Edward H. White floats in zero gravity outside the *Gemini 4* spacecraft on June 3, 1965.

During the Gemini launch on July 18, 1966, eighth in the series of manned space flights, John Young and Michael Collins managed to dock successfully with two different Agena targets. Their successful space walk included transferring between the Gemini and the Agena, and back. But when Richard Gordon tried a space walk from the ninth Gemini flight, he too became exhausted. The learning phase finally ended with the launch of *Gemini 12* on November 11, 1966. This 10th manned flight docked again with an Agena, and included the most successful space walk yet. Astronaut Buzz Aldrin managed to use a set of tools to work on panels on the outside of the Gemini and on the inside of the docking adaptor. By pacing his efforts, he avoided the fatigue and difficulty in climbing back aboard the spacecraft that his predecessors experienced.

descent sequence themselves, the men would have to complete another orbit. But this would place the spacecraft off course. When they eventually fired the rocket and the spacecraft came down successfully, they were 2,000 miles (3,220 kilometers) from the planned landing place, stranded in the snow-topped Ural mountains. Rescue services took several hours to find them.

Problems like these showed that more ambitious flights would require new spacecraft. The Voskhod was retired after two launches. The Russians worked on the Soyuz while the Americans kept perfecting their Gemini spacecraft, which they had tested with two unmanned launches in 1964 and 1965.

Their first manned launch took place on March 23, 1965, just five days after *Voskhod 2*. Besides holding two crewmembers, the Gemini was easier to control than its predecessor, and the crew could practice the techniques and maneuvers needed to dock with other spacecraft. Virgil Grissom and John White stayed in orbit for almost five hours, making three circuits of Earth before splashing down in the ocean. On June 3, 1965, James McDivitt and Ed White stayed in space for four days and 64 orbits. On orbit three, White spent 21 minutes in a space suit on the first American EVA (extra-vehicular activities, or space walk). He also tried out a newly created handheld maneuvering gun.

Clearly, Gemini was a success. When the series ended in 1966, both the Americans and the Russians were developing new spacecraft. The Russian Soyuz would achieve longer and more impressive orbital flights and would perform a variety of more complex experiments. The Americans were also working on a new spacecraft that would carry three astronauts on a much more ambitious voyage that would surpass every previous Soviet achievement. The new spacecraft was the Apollo, and its destination was nothing less than the Moon.

OPPOSITE Photograph of Ed White during extra vehicular activity on *Gemini IV*.

THE FIRST CASUALTIES

Considering all the challenges, early space flights boasted an excellent safety record. Yet disaster struck at the time when both the Americans and Russians seemed to have finished the early experimental phase. In the first case, involving three American astronauts, the spacecraft was not even in flight.

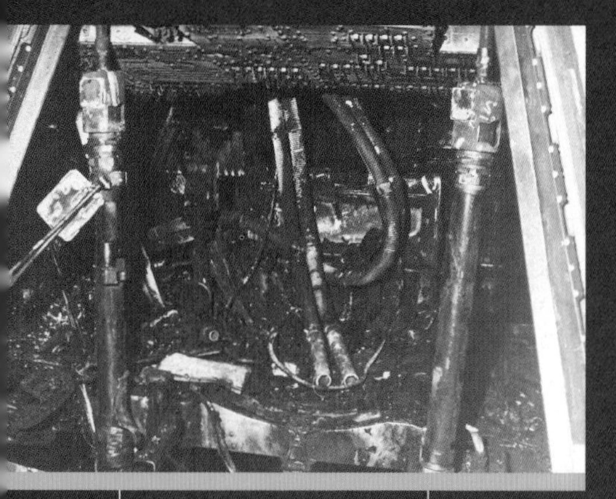

ABOVE Closeup view of the interior of the Apollo spacecraft showing the effects of the intense heat of the flash fire that killed the crew members.

On January 27, 1967, Virgil Grissom, Ed White and Roger Chaffee were aboard the three-man Apollo capsule and breathing pure oxygen as they conducted a series of checks. At 6:30 p.m., instruments outside the capsule showed a surge in the crew's use of oxygen. Another monitor showed that Ed White's heartbeat rate was rising dramatically. Within a minute the crew reported a fire inside the capsule and called to be let out.

Ironically, some improvements that had been made to the hatch design years earlier made this nearly impossible. Instead of the original release handles, they had to unfasten a set of bolts from inside the capsule and that would take more than a minute and a half. Before the crew could do this, the interior was ablaze. The exterior skin became so hot that the emergency services took six minutes to force the hatch open. By then all three astronauts were dead.

Investigations showed that the capsule had three fatal flaws in its design and construction. First of all, the miles of electrical wiring had been badly fitted, with harnesses twisted round corners or left across surfaces where the wire was prone to damage. Second, the plastic materials inside the capsule,

ABOVE Portrait of Virgil Grissom, Edward White, and Roger Chaffee wearing their NASA space overalls.

including netting and straps to hold objects under weightless conditions, were not fire resistant. Third, the pure oxygen atmosphere would turn the slightest spark into a raging fire.

A stray spark from a defective or damaged section of wiring somewhere under Grissom's seat had probably triggered the blaze. As a result, the capsule was radically redesigned with thousands of changes, a completely new hatch, revised wiring and the use of fire-resistant plastics. This delayed the Apollo program nearly two years, but produced a spacecraft that was much safer for crewmembers. The tragedy had taken the lives of three space pioneers, but at least it prevented a potentially more frightful disaster in the depths of space.

Within three months of the Apollo fire, the Russians also lost one of their space pioneers. Vladimir Komarov was aboard *Soyuz 1*, the prototype of the Soviet Union's next-generation spacecraft, when it was launched on April 23, 1967, on a mission to dock with another spacecraft. On his 18th orbit, Komarov initiated the descent procedure for a return to Earth. It was believed that the lines of the parachute deployed to slow down his space capsule became entangled, and he was killed in the resulting crash. For both contestants it was now clear that space itself could be a deadly place.

5

DESTINATION MOON

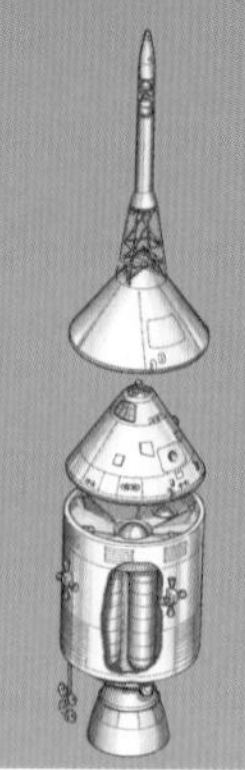

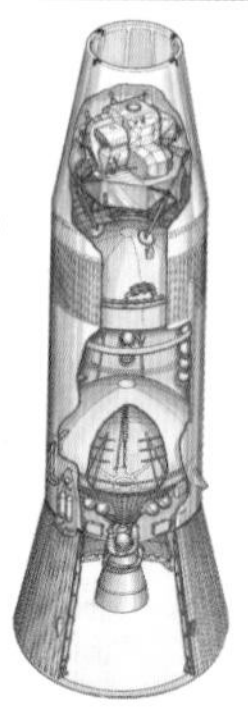

The story of the Apollo missions — the scientific advances, the heroes, the disappointments and those first steps on the surface of the Moon.

President John F. Kennedy took a huge gamble when he announced on May 24, 1961, that America would send a man to the Moon and back by the end of the decade. At that time, the required technology did not even exist. There was no rocket powerful enough to do that job, nor any way to make a controlled landing on the lunar surface, much less take off and return to Earth.

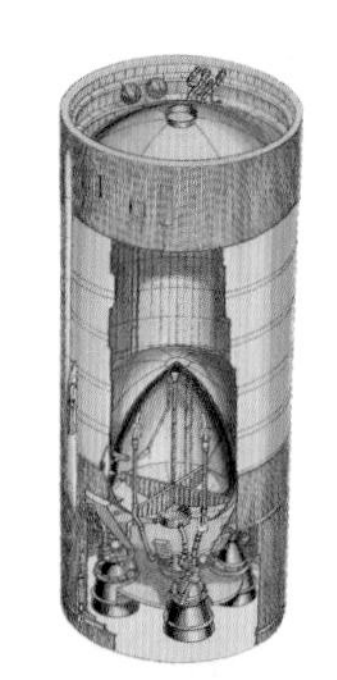

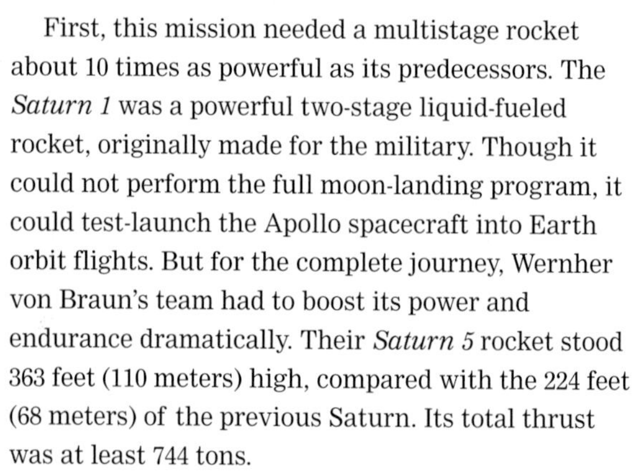

First, this mission needed a multistage rocket about 10 times as powerful as its predecessors. The *Saturn 1* was a powerful two-stage liquid-fueled rocket, originally made for the military. Though it could not perform the full moon-landing program, it could test-launch the Apollo spacecraft into Earth orbit flights. But for the complete journey, Wernher von Braun's team had to boost its power and endurance dramatically. Their *Saturn 5* rocket stood 363 feet (110 meters) high, compared with the 224 feet (68 meters) of the previous Saturn. Its total thrust was at least 744 tons.

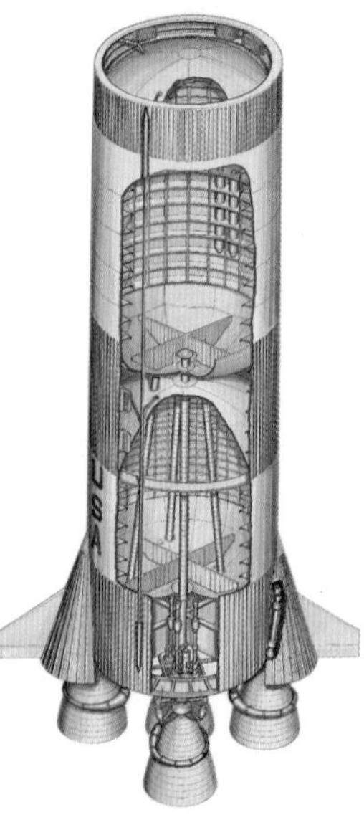

During a lunar mission, each stage of the rocket would perform a critical role. The first stage contained five rocket engines, burning kerosene and liquid oxygen, and strapped together inside a cylinder 33 feet (10 meters) in diameter and 138 feet (42 meters) high. This was enough to lift the payload needed to reach the Moon. Then the first stage would drop off, and the second stage would ignite. This was

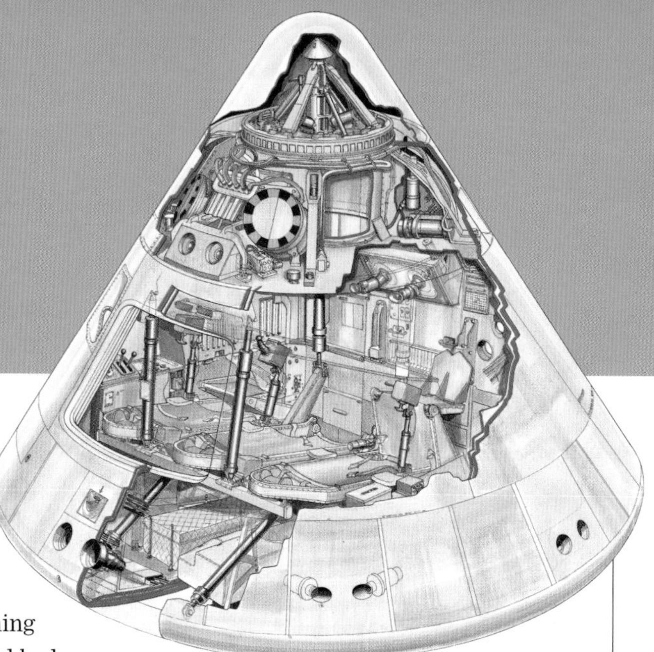

81 feet (25 meters) high, and 33 feet (10 meters) across, with five more rocket engines, this time burning liquid oxygen and liquid hydrogen. It would boost the spacecraft 100 miles (161 kilometers) above Earth before it too would be spent, then drop away to save weight.

Though slightly smaller, the third stage played a much more complex role. It contained a single rocket that burned liquid hydrogen and liquid oxygen. Once the second stage had separated, this third stage would boost the craft into Earth orbit then shut down — the first time a liquid-fueled rocket would be shut down before burning all its fuel. Besides, it would need to be relit and supply exactly the right amount of thrust at the right moment in orbit to accelerate the spacecraft onto its precise trajectory for the Moon.

The spacecraft, which was weightless while in orbit, consisted of several sections, all linked together. The only part that would make the full round trip was the command module, effectively the conical nose of the rocket. It measured 12 feet (3.7 meters) in height and 13 feet (4 meters) in diameter and weighed four tons. Behind this was the cylindrical service module. Behind this again was the lunar module, which would actually land on the Moon's surface.

A successful moon landing would require teamwork and careful timing. When the third stage

ABOVE The command module of the spacecraft.

OPPOSITE The enormous three-stage Saturn rocket.

of *Saturn 5* had been dropped, the lunar module would have to be removed from its position at the tail of the spacecraft. The astronauts would then turn around their command module and service module and redock with the lunar module, at this point fixed to the nose of the conical command module. Once they arrived in lunar orbit, two astronauts would scramble into the lunar module, the hatches would be closed, and the connection severed. While the third astronaut continued orbiting the Moon aboard the command module, the lunar module would descend, using its own engine to slow down and land. The two men would explore the Moon, then reboard the lunar module and fire its ascent engine, leaving its lower descent engine stage behind. The upper stage of the lunar module would climb back into lunar orbit and redock with the command module.

BELOW The lunar module with the lunar rover being deployed.

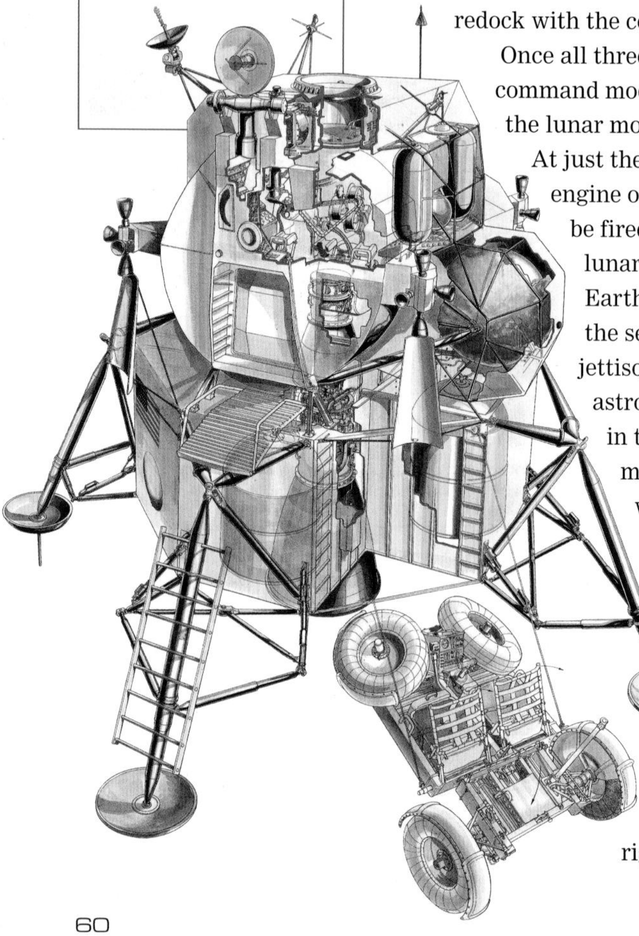

Once all three men were inside the command module, the upper section of the lunar module would be discarded. At just the right moment, the rocket engine on the service module would be fired to speed the craft out of lunar orbit, on a path toward Earth. After it arrived safely, the service module would be jettisoned, and the three astronauts would splash down in the ocean in the command module. To do this they would have to turn the module around. That way, the heat shield on the base could absorb the huge temperature rise caused by friction with Earth's atmosphere during reentry. They also had to hit the atmosphere at the right angle for a safe reentry.

Would these plans work, with so much new technology? Unless the entire complex routine went perfectly, the project could end in disaster. NASA set out to prove that every part of the system would work smoothly. Ironically, this careful test program was the reason why astronauts Gus Grissom, Ed White and Roger Chaffee were on board the first Apollo capsule on January 27, 1967, when it caught fire (see chapter four). After their tragic deaths, the capsule was radically redesigned to make it much safer.

ABOVE The *J2* rocket motor, on exhibition in the United States.

The stages of the *Saturn 5* rocket were tested first. In February 1966, the *Saturn 1B* already had been used to launch the unmanned *Apollo 1* spacecraft into Earth orbit. The huge *Saturn 5* rocket, with the spacecraft aboard, first flew on the morning of November 9, 1967. This grand venture was numbered 4 in the Apollo series and was unmanned, like the previous ones. After *Apollo 4* blasted off from Cape Canaveral in Florida, both the first and second stages of the rocket performed exactly as planned. Finally the third stage fired and put the spacecraft safely into Earth orbit. So far, so good. But the critical phase was yet to come. The new third stage engine would need to be restarted by ground signals since there was no one aboard to do the job. Would it work?

When Mission Control sent the signal, the rocket motor ignited as predicted. The third stage pushed the Apollo spacecraft into a much higher orbit, more than 10,000 miles (16,000 kilometers) from Earth. The spacecraft's own engines also worked as planned. Finally, the unmanned command module made a perfect reentry and splashed down in the Pacific on schedule, just over eight and a half hours after launch.

This brilliant performance showed that men might actually be able to land on the Moon. Unfortunately, one part of the system had not been tested on the *Apollo 4* flight — the lunar module itself. Weight was a critical factor, and the only way

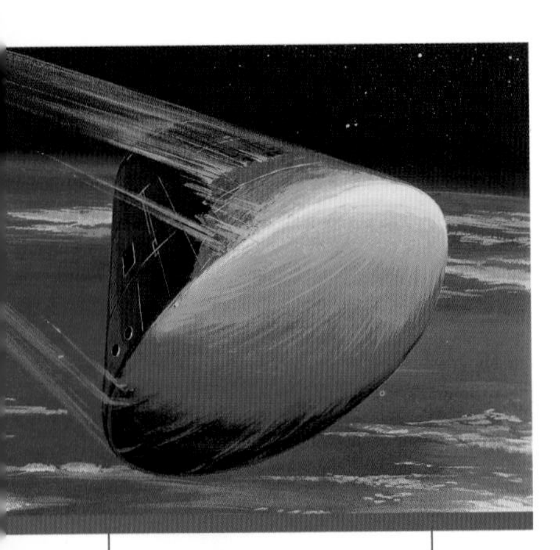

ABOVE An artist's concept of the mission profile of the scheduled *Apollo 8* flight.

to keep the lunar module within weight limits was to design its landing legs to cope with the much weaker Moon gravity. But how could it be tested properly without actually taking it to the Moon?

Instead, *Apollo 5* was used to put the lunar module through its paces in orbit above Earth. The craft was launched on January 22, 1968, and once in orbit, both of its engines were tested. The descent engine was the first rocket engine built to be controllable within very fine limits. The astronauts would have to control the thrust of this crucial engine moment by moment to make a smooth and safe landing. Happily, this engine worked as planned. The ascent engine, which would be the astronauts' only chance to leave the Moon and return home, also performed flawlessly.

Problems arose when they began checking the whole system. *Apollo 6* was launched on a *Saturn 5* rocket on April 4, 1968. The first stage was a success. But the next stage died out prematurely and the third was forced to ignite too early. Then this engine, too, failed to burn long enough to place the spacecraft in Earth orbit. Even worse, it also failed to restart, to deliver the punch needed to put Apollo on course for the Moon.

Engineers were deeply disappointed but also determined to solve the problem. Instrument readings retrieved and analyzed at Mission Control showed that a fuel leak had caused the first premature shutdown. Incorrect wiring had caused the later problems. If these could be corrected, the system should work as well as before.

For the moment, scientists set aside questions about the *Saturn 5*. They focused on the manned test in Earth orbit of the modules that would carry the astronauts to the Moon and back. Since this demanded less thrust than the full-scale Moon mission, the smaller *Saturn 1B* was used to launch *Apollo 7* on October 11, 1968, with three men aboard.

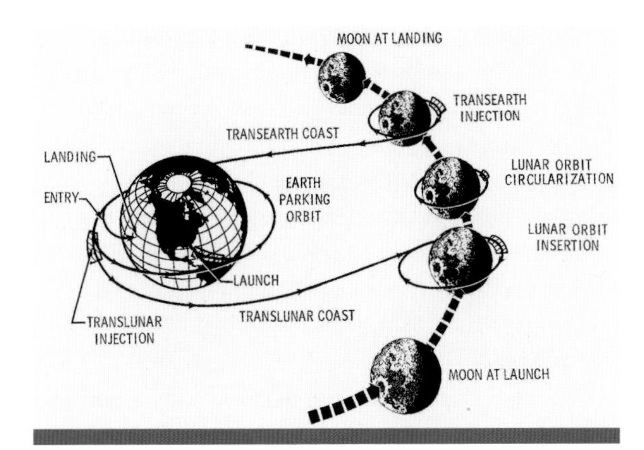

The smaller rocket performed faultlessly, and the module went into Earth orbit for a total of 163 circuits. During this time, the astronauts were able to maneuver their spacecraft to rendezvous with the rocket that had carried them, using the service module's own engine. When they splashed down at the end, only two links in the chain were still unproven: the *Saturn 5* after the correction of earlier flaws and the lunar module with its crew on board.

LEFT TOP An artist's concept of the mission profile of the scheduled *Apollo 8* flight.

BELOW *Apollo 7* crew portrait, taken in May 22, 1968 (From left to right: Donn Eisels, Walter Schirra and Walter Cunningham).

Now plans were made to launch a new *Saturn 5* in December 1968, to prove that every stage would fire properly. By this time, engineers were so confident, they decided to avoid some duplication by combining two steps in the program. When *Saturn 5* was launched three weeks later, it carried *Apollo 8* and astronauts Jim Lovell, Frank Borman and William Anders all the way to the Moon and back. For the first time, a manned space mission had left the confines of Earth behind.

On this historic flight, the rocket worked perfectly, and after one Earth orbit, the men restarted the third stage engine for five minutes, to set a course for the Moon. For safety reasons, this first course would allow them to simply curve around the Moon under the influence of lunar gravity and to automatically return to Earth if any major system failed. Only later, if all went well, would they fire the service module's engine to alter its course to pass within 70 miles (112 kilometers) of the Moon's surface.

Apollo 8 was captured by the Moon's gravitational pull and settled into an elliptical orbit around the Moon. The astronauts turned the craft and used its engine to slow them down. More engine corrections changed the orbit into a more circular path, giving an unparalleled view of the lunar surface. After 10 circuits, they fired the service module's engine again, to head back to Earth. The rest of the flight went well, and *Apollo 8* hit the ocean six days after launch. The future of the lunar mission seemed bright.

Only the manned lunar module test remained, and this was carried out in Earth orbit on a flight that began on March 3, 1969. At a height of almost 120 miles (193 kilometers), the astronauts separated the lunar module from the command module, then turned the command module around and linked it up to the lunar module. Two astronauts climbed into the lunar module, leaving a third man aboard the command module.

LEFT *Apollo 8* liftoff.

The lunar module's engines and control systems were tested before linking up with the command module. The crew also tested new space suits, each carrying its own life-support system. Finally, with all three men back inside the command module, the lunar module was jettisoned, and the command module returned safely on March 13.

Scientists planned a sort of "dress rehearsal" for the whole Apollo program on May 18, 1969. *Apollo 10* was sent even closer to the Moon than *Apollo 8* had done. The three astronauts went through the same routine, but this time the two men inside the lunar module used its descent engine for the first time over the Moon. Slowly, they sank to a height of 8.9 miles (14,300 kilometers) over the planned landing site for *Apollo 11*, on the flat plain of the Sea of Tranquility. Then they climbed away to rejoin the command module and return to Earth.

What would happen when the *Apollo 11* lunar module actually touched down? Would its feet land on solid rock, or soft dust into which it might sink, and endanger its relaunch? NASA planning teams needed to know. A series of Moon probes sent by Russia and the US (see chapter eight) revealed little about the surface. Then in January 1966, the Russians launched *Luna 9,* which made the first soft landing in an area called the Ocean of Storms. It sent back information showing the surface was stable, at least in that spot. The American craft, *Surveyor 1*, made a soft landing on June 2, 1966. The probe's landing feet were designed to place the same pressure on the surface as those of the lunar module. Detailed close-ups showed that only faint marks were left on the surface. Later probes dug into the surface to collect and analyze a soil sample, which confirmed this good news. It was now time for *Apollo 11.*

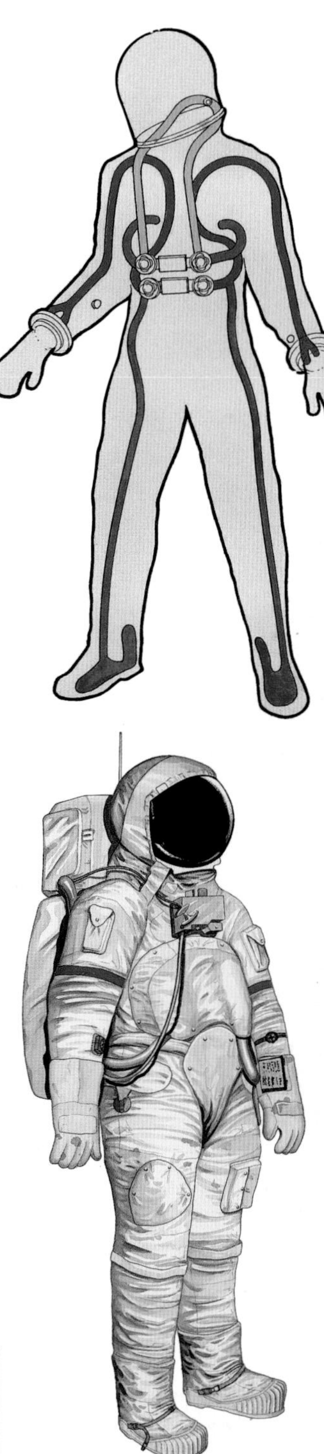

RIGHT The "portable life-support system" for astronauts. Oxygen and power are supplied via the internal cords.

THE EAGLE HAS LANDED

The long-awaited *Apollo 11* mission began on July 16, 1969, with astronauts Neil Armstrong, Edwin "Buzz" Aldrin and Michael Collins aboard. After a successful launch and flight into lunar orbit, Armstrong and Aldrin climbed into the lunar module Eagle and started to descend to the lunar surface. This time they were going all the way.

But some glitches occurred. Communications with Earth were patchy and messages had to be relayed through the command module, Columbia. Huge amounts of information from the lunar module's instruments caused the Mission Control computer system to overload, and a warning light flashed. This seemed like instrument failure rather than a serious problem, and Eagle was told to make an automatically piloted landing.

OPPOSITE ***Apollo 11*** **extravehicular activity on the moon.**

BELOW A close-up view of an astronaut's foot and footprint in the lunar soil.

Then an otherwise small navigation error put the module four miles (6.5 kilometers) from the planned landing site. Armstrong told Mission Control they were approaching a field of boulders, which could cause damage, so he took over the landing manually. Slowly, softly and above all safely, Eagle landed on the lunar surface at 8:17 p.m. Greenwich Mean Time on July 20, 1969.

Armstrong and Aldrin prepared the craft for the return journey, ate and then climbed into their space suits for their first walk on the lunar surface. After carefully descending the ladder and taking his "one small step for man, one giant leap for mankind," Armstrong made history as the first human to reach the Moon. For two hours, the men collected soil and rock

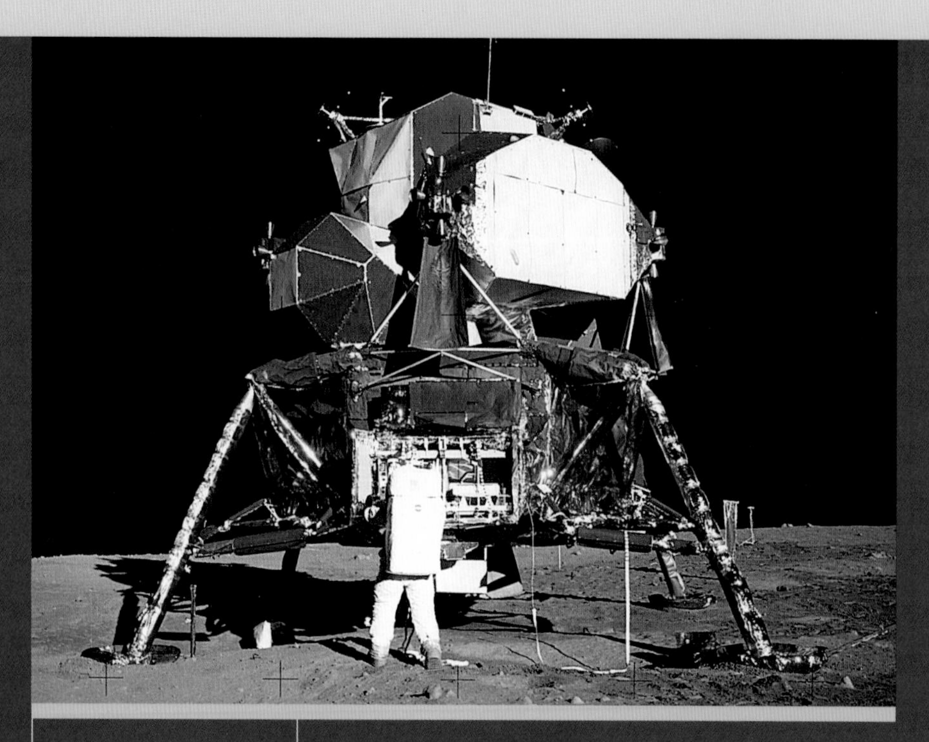

ABOVE Astronaut Edwin E. Aldrin Jr., lunar module pilot, prepares to deploy the Early Apollo Scientific Experiments Package (EASEP) on the surface of the Moon.

OPPOSITE TOP The Moon.

OPPOSITE BOTTOM The three *Apollo 11* crew members await pickup by a helicopter from the USS *Hornet*.

samples. They set up a series of experiments, including equipment for measuring the solar wind and a laser reflector to enable precise measurements of the distance between Earth and the Moon to be made.

After returning to the module, they slept until it was time for the launch. At 6:03 a.m. Greenwich Mean Time on July 21, Eagle soared upward to rendezvous with Columbia. Despite some problems during the final docking, the two craft linked up, the astronauts climbed back into the command module, and the lunar module was released. Their routine flight back to Earth ended with splashdown in the Pacific on July 24, 1969. President Kennedy's bold promise had been kept.

Apollo 12 completed another successful lunar mission in November 1969. Apart from *Apollo 13*, the Apollo flights remained remarkably reliable and safe. After changes were made, *Apollo 14* was launched on January 31, 1971. The lunar module landed in the Fra Mauro area of the Moon. Once again, astronauts set up experiments and collected samples before returning to Earth on February 9, 1971. Later that year, *Apollo 15* went to the Hadley Rille area. This was the first outing for the four-wheeled battery-powered lunar roving vehicle, which greatly increased the crew's range.

Apollo 16 took astronauts to the Moon's Descartes Highlands. Finally, on December 7, 1972, *Apollo 17* headed for the edge of the Sea of Serenity, near the crater Littrow. This was the last flight in this amazing series. It broke all the previous records for the time spent on the Moon, the distance traveled by lunar rover and the retrieval of more than 250 pounds (113 kilograms) of samples.

ADRIFT IN SPACE — UNLUCKY 13

Following the careful test program and outstanding success of *Apollo 11* and *Apollo 12*, the launch of *Apollo 13* did not seem unusual. All went well as astronauts Jim Lovell, John Swigert and Fred Haise headed toward the Moon. But disaster struck just past the midcourse correction point. A liquid-oxygen tank in the service module exploded, and it was soon clear that the spacecraft might run out of power. One tank had failed and the other showed signs of failing. The only other power source was the batteries aboard the command module to be used during reentry.

ABOVE The severely damaged *Apollo 13* service module taken by a Maurer motion picture camera from the lunar module.

The implications were terrifying. Power was needed for heating, light, life support, instruments and all the other tasks to keep the astronauts and the spacecraft alive. The spacecraft must be brought back to Earth as soon as possible. But turning it around and reversing its course would require too much power. The most practical choice seemed to be to correct its course so that it could swing around the Moon before heading home. Yet, the power and life-support systems in the service module could not last that long.

The astronauts seemed doomed unless they could use the lunar module as a space lifeboat. Since this module had its own power and life-support systems for the lunar landing, they could survive there much longer. To boost the water supplies, the men used plastic bottles to bail water from the command module tank. They re-rigged air supply connections with plastic sheet and adhesive tape. Meanwhile, they saved battery power by using the lunar module engines to carry out the maneuvers that would correct their course on the way back to Earth.

By the time they approached reentry, the men were cold, exhausted and dehydrated. When they dropped the service module, they were shocked by the size of the hole blown in the side by the exploding tank. Then they powered up the command module batteries and saw that the current was draining out far too quickly.

Only when communications were restored and the circuits turned on did they find that a switch had been left on. The crew transferred to the command module, and the lunar module that had saved their lives was finally jettisoned, to drop into the ocean. Their final fear was the trajectory taken up by the spacecraft, which was perilously close to bouncing away into space. Mission Control found the error had been caused by water boiling off the cooling system on the lunar module. Once that module was dropped, the service module remained barely within the limits for a successful reentry and a splashdown that brought cheers and tears of relief from those on the ground. The Apollo safety record had held, but just barely.

ABOVE **Recovery of the *Apollo 13* astronauts, aided by underwater demolition team swimmers.**

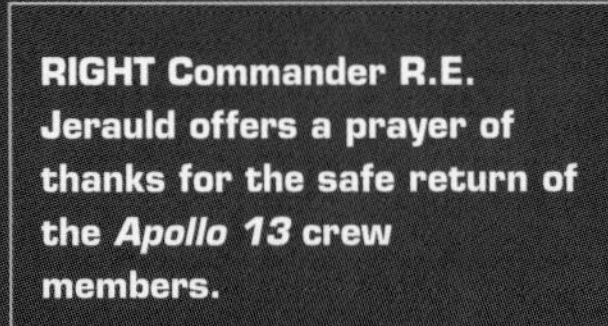

RIGHT Commander R.E. Jerauld offers a prayer of thanks for the safe return of the *Apollo 13* crew members.

6
THE SPACE SHUTTLE

The launch of the space shuttle brought a new era in space travel and laid the foundations for long-term space exploration.

After the stunning success of the Moon landings, America faced a big question: Where should NASA take the space program now? The planet Mars seemed like a logical place to explore next. But landing people on its surface would be too costly.

RIGHT The space shuttle at the start of its near-vertical climb after launch.

Maintaining a program of unmanned probes would be cheaper and easier, and would teach us more about the solar system, and the wider universe. Through a series of flights, NASA could build a space station. From there, they could conduct valuable long-term experiments and possibly lay the foundations of future longer-distance space flights.

But certain obstacles stood in the way. The Saturn rocket used for the Moon launches was terribly expensive. Most of each rocket, which cost US$120 million at the time, was discarded after one use. NASA could save vast amounts of money by creating a reusable space launch vehicle. It could be brought back to Earth after each mission to be refueled and serviced, and then used again, as often as necessary.

Before NASA could solve the technology questions, they faced other problems. The US Senate rejected the space station project. This meant there was no need for a reusable manned launch vehicle, since future space research would use only unmanned satellites and space probes. Yet satellites played an important role in the process that would eventually produce the space shuttle. The shuttle would reduce the cost of sending human astronauts into orbit and make satellite launches cheaper and more reliable. It would even make it possible to service and repair existing satellites.

In some ways, developing a shuttle vehicle was more difficult than producing a powerful and dependable rocket. By the late 1960s, rocket technology was well advanced, with accurate control systems that worked within the atmosphere and in

ABOVE Artist's concept of the space shuttle as it enters Earth's atmosphere.

BELOW Launching the space shuttle *Discovery* on June 2, 1998.

THE OTHER SPACE SHUTTLE

The Soviet version of the space shuttle clearly showed that the Americans had designed their craft and its supporting systems well. In 1978, Russian scientists and engineers began working on a project coded VKK (the initials stood for the Russian equivalent of Air-Space Vehicle System), which would consist of the huge *Energiya* rocket, and a reusable shuttle-type hybrid space vehicle/aircraft.

Their set-up looked like a twin of NASA's shuttle and its main rocket, though some details were different. The Russian shuttlecraft, named *Buran*, had maneuvering engines but no main engine. Instead, the main *Energiya* booster, with its thrust of 3,500 tons, would provide the power. Besides, there were no separate twin boosters and main tank, since the rocket combined both functions with four strap-on boosters and a second stage that were jettisoned during the ascent.

OPPOSITE ***Buran*** **being carried by** ***Energiya*** **in flight, May 19, 1989.**

BELOW The ***Energiya*** **booster rocket with** ***Buran*** **on the launch pad.**

During the first test launch on May 15, 1987, damage resulted from the powerful rocket's own blast as *Energiya* lifted off. To solve that problem, a concrete flame pit five stories deep was dug in the center of

the launchpad to absorb the rocket exhaust and carry it away through vents. The crews boarded the *Buran* through a tubular walkway from a protected bunker. In an emergency they could slide down an escape chute into the bunker in a matter of seconds.

The *Buran*'s first flight took place on November 15, 1988, but was operated without a crew, under automatic control. The *Energiya* rocket lifted the shuttle into initial Earth orbit at a height of 99 miles (160 kilometers). The shuttle's own maneuvering engines boosted the craft to a height of 155 miles (250 kilometers) for two complete orbits. *Buran* was then turned around and the maneuvering engine fired to slow down the craft and begin its return to Earth.

Still under automatic control, the *Buran* touched down on a special airstrip close to the Baikonur launch site. It had flown for three hours 25 minutes. The Soviets announced they would conduct four shuttle flights a year. But they seemed to drop that plan in favor of conventional rocket launches, while planning that *Buran* and its successors would play a key part in any mission to send cosmonauts to Mars.

space. However, now researchers were trying to make a space aircraft that could fly in the ordinary way once it re-entered the Earth's atmosphere.

Step by step, the basic design of the space shuttle unfolded. NASA first considered using a rocket to launch the shuttle into orbit, but with a pilot on board to return the rocket to Earth. This idea proved to be a blind alley. By 1971, they had decided that the only piloted part of the shuttle would be the orbiter. To launch this into orbit around Earth would call for a pair of solid-fuel rocket boosters. They would drop off when their fuel was spent and would carry parachutes that would let them drop back to the surface slowly so that they could be retrieved and reused for other launches.

The orbiting shuttle would continue to rise on its three main engines, supplied with liquid oxygen and liquid hydrogen from a huge fuel tank strapped

BELOW The successful capture of the *Intelsat VI* satellite is recorded over Mexico from the shuttle's *Endeavor* cabin.

below the orbiter. Once the shuttle approached its orbiting height, the main tank would be jettisoned, breaking up as it fell, later hitting the ocean. This was the only major part of the shuttle that would not be reusable. For the rest of its flight, the shuttle would use its own engines, which would also be used to slow down the craft when it returned to Earth.

Perhaps the greatest challenge was making sure the shuttle could survive re-entry into the atmosphere and remain controllable right through to the landing. One vulnerable area was the wings. The US had tried out a high-altitude research aircraft with no wings at all, but it could not be controlled at reasonable speeds in the lower, denser layers of the atmosphere. Wings would be needed. These had to be as large as possible to provide enough control for a safe landing. But over-large wings would be weak at re-entry, and might even be torn off when they re-entered the atmosphere.

In the end, the small delta wings and tailfin fitted at the rear end of the shuttle proved to be strong enough during re-entry. They let the pilot make an accurate approach and then touch-down at a blistering 200 miles per hour (322 kilometers per

ABOVE Astronaut Bruce McCandless II is pictured on the manned maneuvering unit above earth on February 7, 1984.

BELOW A front view showing *Columbia*'s touchdown at Kennedy Space Center's landing facility.

ABOVE Astronaut Bruce McCandless II, conducts a simulation of a chore scheduled on a later mission to aid the damaged Solar Maximum satellite.

hour), on a special runway at Edwards Air Force Base in California's Mojave Desert.

Surviving the soaring temperatures of re-entry proved more difficult. The craft needed top-quality protection, especially around key areas, like the leading edges of the stub wings and the nosecap of the orbiter. Specially made black insulating tiles were layered on the underside of the craft. They could withstand temperatures around 2,172 degrees Fahrenheit (1,189 degrees celsius). Similar white tiles were attached to the shuttle's upper surfaces for lower temperature insulation, while reinforced carbon composite tiles were fixed to the high-stress areas on the wings. The tile material was reliable, but the curves and edges of the craft's surface caused a problem for the adhesive used to attach them, and they often dropped off. NASA worked on the problem while the rest of the program was being completed.

In 1981, the first space shuttle, *Columbia,* soared into history. It was a curious blend of space vehicle and conventional aircraft. The flight-deck, situated in the upper section of the nose, has two seats set side-by-side, for the mission commander and the pilot. Behind them are two more seats for a mission specialist and the astronaut in charge of the payload. Below the flight deck is a mid-deck that can seat up to three more passengers, along with crew bunks and a galley. A central air lock gives the crew access to the unpressurized central cargo bay, which carries the main mission payload. Below the mid-deck, a lower deck is used as an equipment bay to hold lockers, space suits and life support systems.

BELOW *Discovery's* cargo bay with the air lock clearly open nearby.

THE SHUTTLE AT WORK

To enable the shuttle to launch satellites into higher, geostationary orbits, scientists developed new engines. They attached to the satellites, to propel them from the cargo bay to the right height. First came the 20-ton Interim Upper Stage engine (IUS) which could lift a 2.5-ton satellite into geostationary orbit. Different Payload Assist Modules were developed to do the job for smaller satellites.

The two-deck pressurized cabin of the orbiter was too small for certain experiments. In response, the European Space Agency developed the Spacelab project, a form of semi–space station built up of different modules. These modules were designed to fit within the shuttle's cargo bay and could be used for many different experiments.

TOP The cargo bay of the space shuttle *Challenger*.

ABOVE A shuttle astronaut in space in 1984, riding on the shuttle's Canadian robot arm also called the canadarm.

When the fifth shuttle was launched in November 1982, the craft was ready to handle a commercial project. Once established in orbit, it boosted two communications satellites into higher orbits, as planned. Unfortunately, events did not always go smoothly. In April 1983, the IUS launching NASA's own Tracking and Data Relay Satellite (TDRS) for space communications broke down. It left the satellite at too low an altitude, and ground controllers had to issue careful commands to the satellite's own stabilization thrusters to successfully push it to the required height.

Success came more easily in April 1984. The 11th mission in the series was mounted to repair the Solar Maximum Mission satellite, an Earth-orbiting space observatory launched in 1980. With the controllable robot arm of the orbiter, the crew captured the satellite and transferred it into the cargo bay, then made repairs before relaunching it into orbit. Later that year, the astronauts made use of a jet-powered backpack called the Mobile Maneuvering Unit (MMU). It let them move away from the orbiter, under their own power, to a maximum range of some 300 feet (91 meters), without any line to tether them to the craft.

More recently, the shuttle was used to repair the *Hubble Space Telescope,* which originally suffered from blurred images. Launched on December 2, 1993, the shuttle *Endeavor* made a successful rendezvous with the telescope. The astronauts installed an ingenious ten mirror array to correct the vision of the *Hubble*'s main mirror, as well as a new wide-field planetary camera. They also replaced the telescope's faulty gyroscopes and worn solar panels. Afterward, the telescope began to produce stunning images of the previously unseen older and more distant parts of the universe.

ABOVE The freedom of space — floating above the shuttle's cargo deck, against a backdrop of the ocean.

BELOW Astronaut Gregory J Harbaugh, mission specialist, floats horizontally in the cargo bay of the space shuttle *Discovery*.

Critics point out that launching satellites with the shuttle actually costs more than using once-only rockets. But the shuttle project made possible a wide variety of experiments and missions that would have been impossible otherwise. Significantly, scientists have been working on several new low-cost launch systems. One example is the reusable launch vehicle (RLV). It would take off vertically and land horizontally like the shuttle, but with three key differences. The RLV would be unmanned, it would use metallic heat shields instead of the silica tiles of the shuttle, and it would carry its boosters and fuel tanks with it instead of dropping them during the climb into orbit.

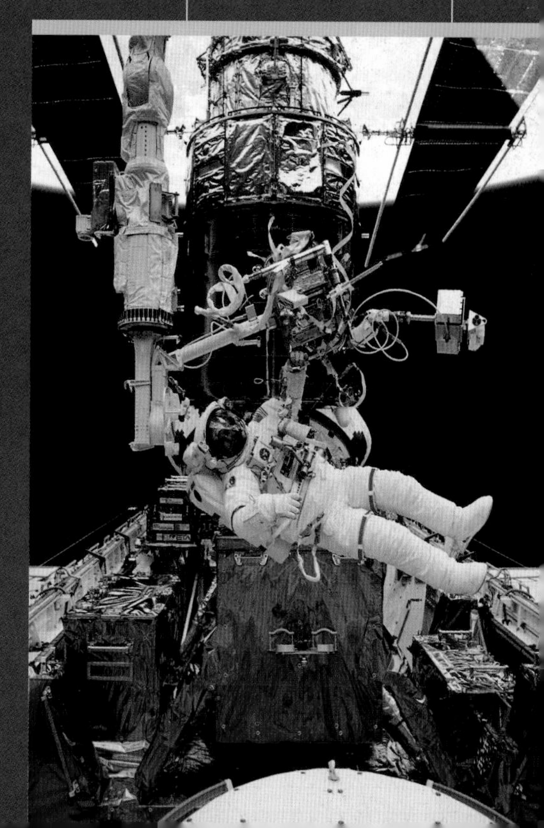

ABOVE Almost routine — space shuttle launch, 1980s.

The key part of the craft is the cargo bay, since this is the real reason for the entire project. The shuttle can carry up to four communications satellites, each with its own booster to fire it even higher, to the altitude needed for a geosynchronous orbit. It is also fitted with a controllable arm, which can be used to retrieve existing satellites from orbit and place them inside for repair.

The shuttle's prospects seemed enticing, but various malfunctions delayed the launch schedule. Finally, on the morning of April 14, 1981, the countdown began. With exactly 3.8 seconds to go, the onboard computers triggered the ignition of the first main engine. Then engines two and three were lit, at intervals of about one-eighth of a second each, and the thrust delivered by these three engines increased to the point where they were lifting the 2,000 ton *Columbia* off the launchpad. At that point, the launch control system fired the two solid-fuel boosters, and the bolts holding the whole assembly to the pad were ignited. The huge shuttle assembly blasted upward, driven by the thrust of five rocket engines.

Just five seconds after liftoff, *Columbia* began a complex series of maneuvers. It rolled through 120 degrees to place the huge fuel tank uppermost. After two minutes and 12 seconds the two boosters detached and fell away, their fuel exhausted. Six minutes later and *Columbia* reached the peak of its first climb, at an altitude of 81 miles (130 kilometers). It then descended to 72 miles (116 kilometers), where the main fuel tank was dropped safely. Then *Columbia*'s maneuvering engines were restarted to boost the craft into an operating orbit — about 170 miles (273 kilometers) above the Earth's surface.

Later on, it would reach altitudes of between 115 and 690 miles (185 and 1,110 kilometers).

During *Columbia*'s historic first flight, the time in orbit was used to check the craft's operating systems, and particularly the tiles and the cargo bay doors. Finally, the crucial pre-entry maneuvers were carried out. First the shuttle had to be turned around so that the main maneuvering engine could be fired to slow it down and drop it out of orbit over the Pacific. Then *Columbia* had to be turned around again into the correct attitude for re-entry, facing forward in the direction of flight, and the right way up.

Before landing, pilot John Young flew the shuttle like a huge glider, carrying out a series of S-turns to lose speed and height. The craft had to be landed right the first time, since the power and control systems did not allow for a second landing. Young had *Columbia* lined up on the glide path well in advance. He proceeded to fly the craft exactly on trajectory until it was almost touching the runway. A slight back pressure on the control column lifted *Columbia*'s nose in a perfect flare, and the craft settled back onto its main landing gear at 200 miles per hour (320 kilometers per hour). As the pilot braked to a stop, it was clear that the future of manned space flight was now "shuttle shaped."

BELOW The shuttle landing sequence, which finishes with the deployment of the drag chute.

TRAGIC SETBACKS

On February 1 2003, the *Columbia* space shuttle was returning from another routine space mission. However, as it entered Earth's atmosphere disaster struck. As onlookers watched in horror the shuttle disintegrated before their very eyes. Just as the world was beginning to take space travel for granted, this served as an ugly reminder of the dangers of human space exploration. Not since the *Challenger* explosion in 1986 had such a tragedy occurred.

Ground controllers frantically tried to make contact with the *Columbia* crew when they first noticed that the shuttle was having difficulty correcting its attitude upon its re-entry into Earth's atmosphere. They lost all contact with the shuttle as it soared above Texas, traveling at 18 times the speed of sound and at an altitude of 40 miles above the Earth.

Witnesses described hearing a huge bang and seeing several white trails in the blue sky. As they looked on in horror the shuttle broke up, taking with it the seven astronauts on board. While the public tried to make sense of what had happened, NASA quickly announced that independent experts would conduct a thorough investigation.

All photographs, video footage and debris were collected. Much of the debris was badly damaged by the intense heat of the explosion making it difficult for investigators to analyze. The Columbia Accident Investigation Board released its report in August 2003 and concluded that foam insulation that had struck the shuttle during its launch caused the disaster. Columbia suffered a devastating puncture that allowed super-heated air, or plasma, to penetrate deep inside the left wing during re-entry, effectively melting it. Unable to withstand this heat, the shuttle disintegrated.

After the *Challenger* explosion in 1986 further space missions were postponed for two years. Seventeen years later the *Columbia* tragedy revived debates over the future of space exploration. Are the economic and human costs too high? Should we launch only unmanned probes in the future? Despite this, NASA has revealed plans for an Orbital Space Plane — the series of space vehicle expected to replace the shuttle from 2012. This new craft's primary function will be to ferry crews to and from the *International Space Station* (ISS) and serve as a lifeboat if the station has to be evacuated. Space research will inevitably continue and the future of human space exploration will depend on the bravery of astronauts who are fearless enough to venture into the unknown.

ABOVE A traditional portrait of the STS-107 *Columbia* crew members. Seated in front, from left, are astronauts Rick D. Husband, Kalpana Chawla and William C. McCool. Standing behind from left are David M. Brown, Laurel B. Clark, Michael P. Anderson and Ilan Ramon.

OPPOSITE An expanding ball of gas from *Challenger*'s external tank creates a plume of white cloud in the sky.

7
SPACE STATIONS

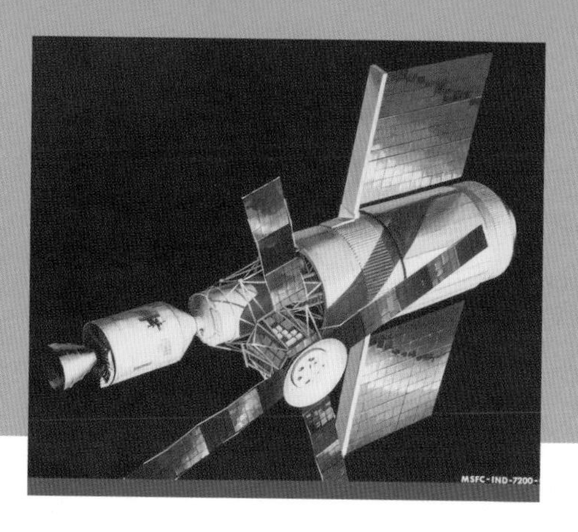

From the unmanned Orbiting Solar Observatories to the Mir space station, the ability to explore space from an outpost has greatly increased our knowledge of the universe.

ABOVE The Apollo Telescope Mount (ATM) is featured in this artist's concept of the Skylab cluster.

OPPOSITE An artist's concept of an American Apollo spacecraft rendezvousing with a Soviet Soyuz in Earth orbit.

In 1962, America launched the first of a series of orbiting solar observatories. These functioned as unmanned space stations that monitored solar activity. By 1975, eight of these orbiting observatories kept a close watch on the Sun and sent back lots of information. Still more information came from a series of orbiting geophysical observatories, orbiting astronomical observatories and the Soviet elektron, proton and prognoz series of satellites, launched from 1964 into the late 1980s.

The first manned space stations appeared after the Apollo moon landings. The Russians scored this coup by launching their *Salyut 1* on April 19, 1971, without a crew. Its maneuvering engine was mounted at the lower end, with a docking port at the upper end. A pair of solar panels at each end opened out to generate power once the craft was safely established in orbit.

Four days later *Soyuz 10* carried three cosmonauts into orbit to test the new system, linking the hydraulic and electrical circuits of both craft as part of the docking process. After two days in space, the two craft separated and the *Soyuz* returned to Earth. The *Salyut* remained in orbit, awaiting the next crew.

On June 6, 1971, cosmonauts Georgi Dobrovolsky, Vladislav Volkov and Viktor Patsayev drew near the space station, under automatic control, aboard *Soyuz 11*. When they came within 330 feet (100 meters) of

APOLLO-SOYUZ
and the International Space Station

As relations improved between Russia and the United States, scientists working on space programs in both countries shared more information. One valuable area of cooperation was the development of a compatible docking system to rescue troubled spacecraft. This project began in 1970, and, in July 1975, an Apollo spacecraft docked in Earth orbit with a Soyuz. Astronauts moved between two spacecraft and conducted joint experiments.

This operation was far more complex than it appeared. Various aspects of the American and Russian spacecraft had to be changed to make them more compatible, along with the docking fittings and procedures. This careful work helped pave the way for the more ambitious joint missions on the Mir space station, and the development of the even more ambitious International Space Station.

ABOVE Soviet Soyuz spacecraft photographed from a rendezvous window of the American Apollo in Earth orbit.

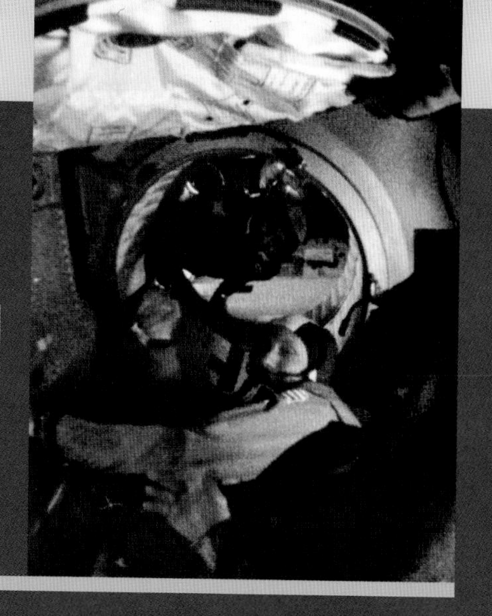

RIGHT Astronaut Thomas P. Stafford (front) and cosmonaut Aleksei A. Leonov make their historic handshake in space on July 17, 1975.

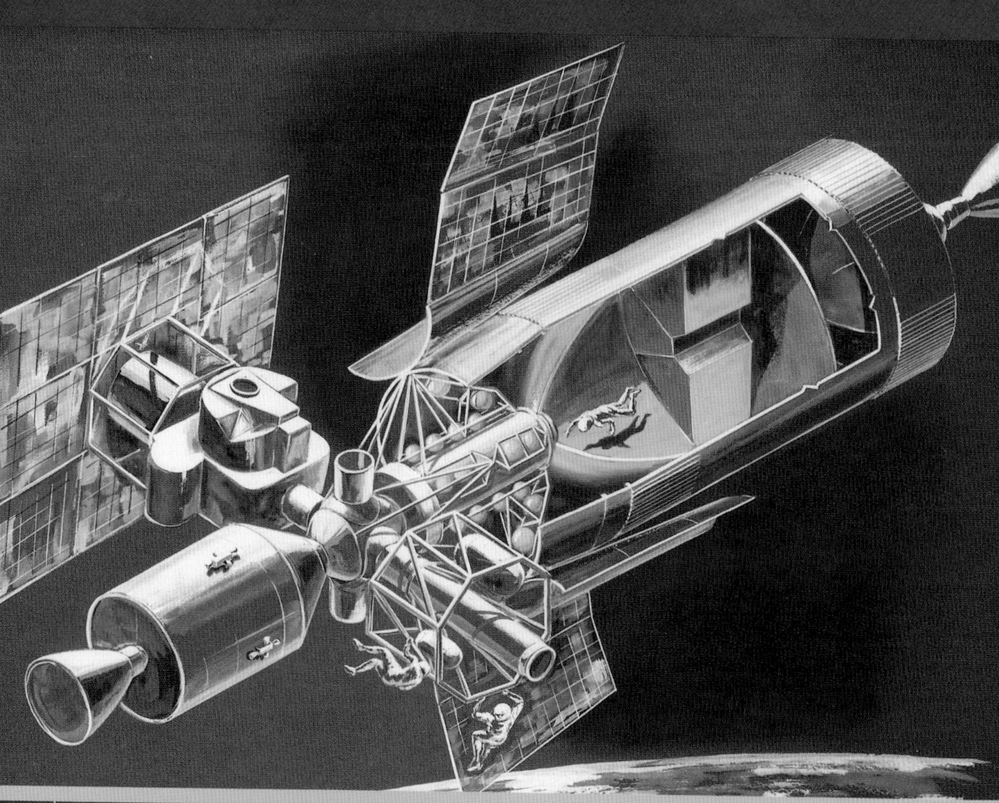

ABOVE An illustration of a future concept for a space station.

ABOVE Junior researcher Y.G. Pobrov observes testing of the Apollo-Soyuz docking system at Rockwell International's plant in Downey, California.

Salyut, the men began using manual control until the two craft made contact. After the docking was completed, they wriggled through the connecting tunnel to board the station. There they carried out a range of experiments, including a series of plant-growth trials to develop food for long space flights.

After completing their program, they re-boarded *Soyuz* to return to Earth. But disaster struck during this most routine part of the mission. Space inside the capsule was limited, so the crew could not wear space suits. Unfortunately, during the re-entry into Earth's atmosphere, the pressure equalizing valve failed, allowing the cabin air to escape into space. It landed perfectly, but all three cosmonauts were dead.

On October 11, 1971, after 175 days in orbit, Soviet Mission Control turned *Salyut 1* around and fired its maneuvering engine to slow it down. This caused it to lose altitude and finally break up on re-entry over the Pacific. The first manned space station had been both promising and disappointing.

Salyut 2 was short-lived. Just minutes after it was launched on April 3, 1973, American tracking stations picked up signals showing fragments had dropped away from the rocket lifting it into orbit. Within three weeks, the project was abandoned.

Once again, the space race was heating up. Before the Russians could launch their next space station, the Americans launched Skylab on May 14, 1973. The station itself was constructed, at minimum cost, from the fourth stage of a *Saturn 5* rocket. The cylindrical liquid hydrogen tank was converted into a roomy two-level accommodation for a crew of three. The lower level contained sleeping compartments, and the upper level held a spacious workshop and laboratory.

Like the Salyuts, Skylab was launched unmanned. A crew of three was to follow aboard an Apollo command and service module combination. However, the station had suffered serious damage shortly after launch. Air pressure had torn away the protective shield that repelled sunlight from the large window of the workshop, along with some of the solar panels that would provide power for the station. Without the shield, the temperature inside the station would soar and the power supplies would decline. Mission Control found a temporary solution. They steered Skylab into an altitude that partly shaded its interior from overheating while keeping the ATM solar arrays in a position where they could produce enough power to run the station.

Astronauts Charles "Pete" Conrad, Joseph Kerwin and Paul Weitz were launched into orbit on May 25, 1973, to repair Skylab. More than seven hours later, they reached their target. While checking the damage, they found that one solar array was missing, and the broken shield had prevented its twin from moving into position. They also found malfunctions in the docking system.

One by one, the problems were solved. The men modified the docking system switches so they could connect their spacecraft to the Skylab entry port. Once they entered Skylab, they found that all was well, but it was too hot to sleep on board. Instead, they slept in the the multiple docking adapter which had its own heat shield.

To lower the heat inside Skylab, they put on space suits and climbed out through an air lock to deploy a makeshift sunshade. This caused the internal temperature to drop so the station was habitable. During their month in space, the three astronauts managed to cut away the obstruction so that the remaining solar array could deploy properly and charge the station's storage batteries. They also took pictures of Earth's surface and of solar activity, and studied the effects of weightlessness.

The second crew, launched on July 28, 1973,

BELOW An illustration showing a cutaway view of the Skylab 1 Orbital Workshop (OWS), one of the five major components of the space station.

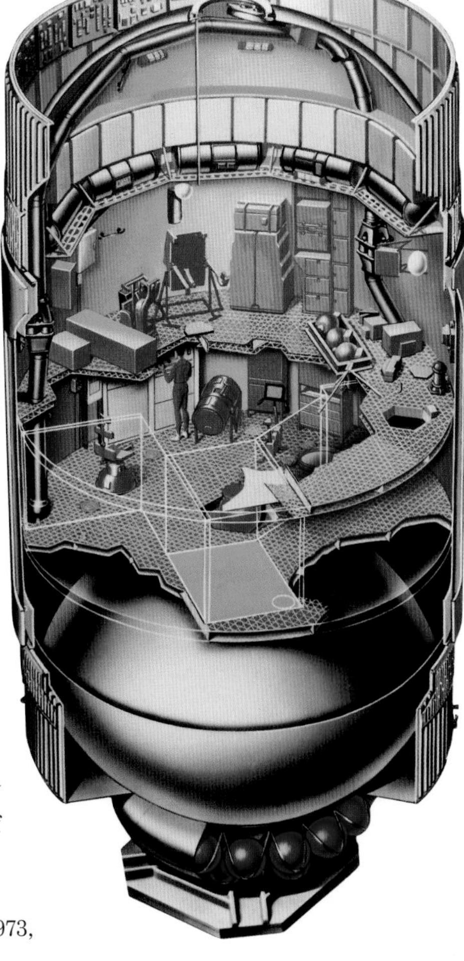

LIVING IN ZERO GRAVITY

As space stations made it possible for astronauts to live for much longer periods in weightless conditions, designers looked for ways to keep them comfortable. Aboard Skylab, each crewmember slept in a vertical sleeping bag and could use a shower and toilet compartment designed to function in zero gravity. A hosepipe with a suction head removed the floating droplets of water that remained after washing or showering. Foods were chosen for variety and ease of eating, and the station included heating and cooling equipment.

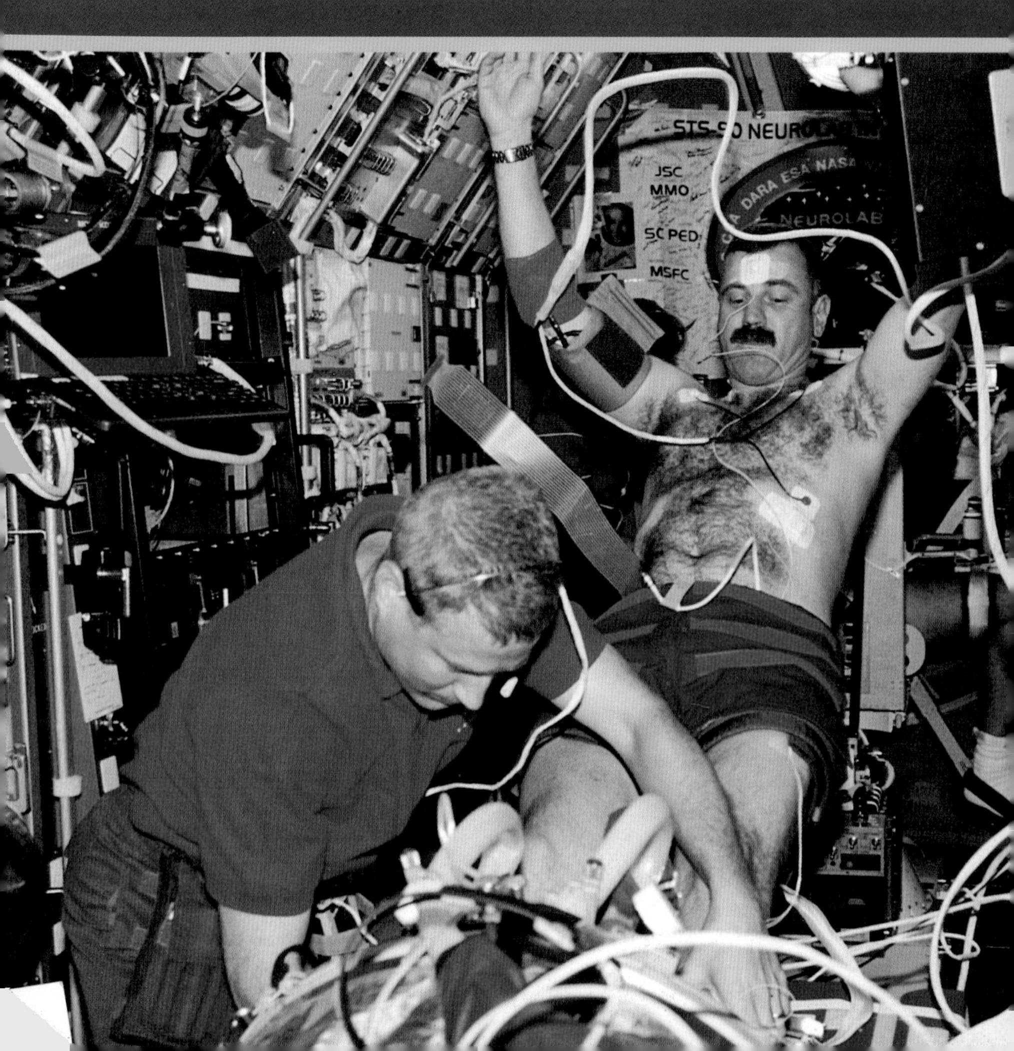

Aboard Mir, astronauts wanted to have one surface of the operations area designated as the floor, despite the lack of gravity. This "floor" was covered in dark green carpet, with light green walls, and a ceiling lit with fluorescent lamps, while the living area was painted in relaxing soft pastel shades.

Both American and Russian astronauts experienced problems adjusting to weightless conditions. On Earth, gravity causes blood and other body fluids to flow down toward the legs, from which they are pumped upward by the heart. Without gravity, these bodily fluids flowed into the upper parts of the body. This depressed their appetite, so the astronauts tended to eat little, but often. To lessen these effects, Skylab carried an exercise bicycle and a negative pressure chamber that expanded the body's lower blood vessels to improve circulation. The Russians used a treadmill for exercise and a clever spring-loaded sweatshirt to exercise the upper body muscles while running.

ABOVE Oranges and grapefruit brought from Earth get a popular reception by the Mir-22 crew members.

OPPOSITE Dafydd R. Williams from the Canadian Space Agency completes a "run" in the Lower Body Negative Pressure device onboard the Spacelab Science Module in 1998.

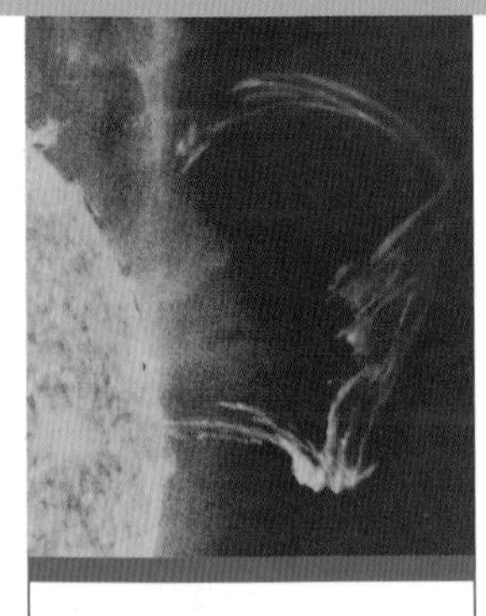

ABOVE A huge solar eruption seen in this enlarged spectroheliogram obtained during the *Skylab 3* mission that took place from July to September 1973.

BELOW A color density rendition of the solar eruption taken from *Skylab 3* in 1973.

spent 59 days aboard Skylab, as planned. The third and final crew reached the station on November 16, 1973. During their 12 weeks in space, the effects of weightlessness, plus a strict exercise program, caused each crew member to grow an inch or more in height as their spines stretched in the absence of gravity.

No further Skylab missions were scheduled. Instead, NASA planned a space shuttle mission in the early 1980s to attach a powered module that could be used to boost the station into a higher and safer orbit. In 1978 and 1979, however, increased sunspot activity caused Earth's atmosphere to expand, exerting additional drag on the empty station. Within weeks of April 1979, Skylab plunged to Earth, landing in the Indian Ocean.

After Skylab fell, it was Russia's turn to forge ahead. They used a different approach to produce larger stations. They developed automatic docking systems and multiple docking ports that allowed space stations to be extended and resupplied for longer periods.

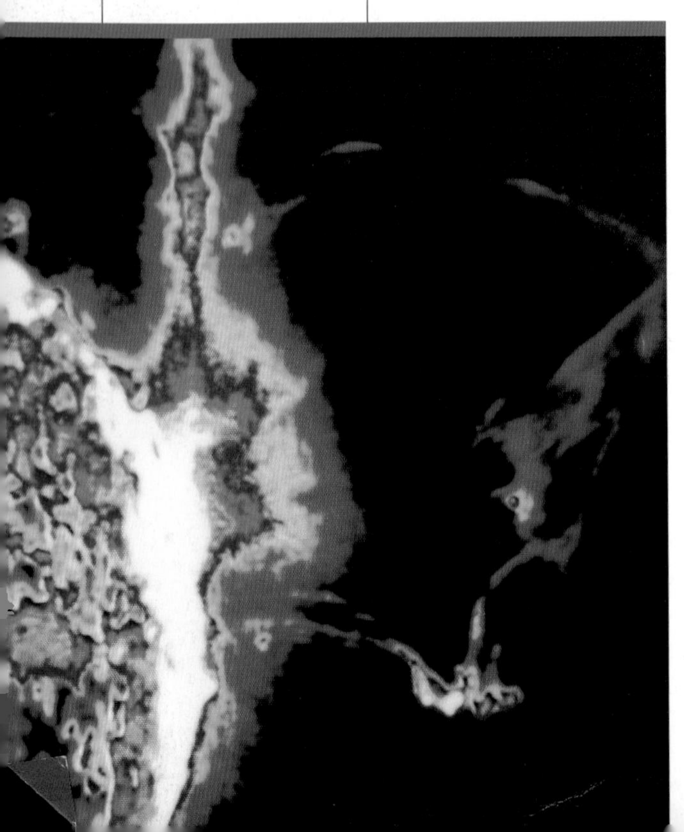

Salyut 3 had been launched on June 25, 1974, and eight days later two cosmonauts were sent up. They stayed on board for just over two weeks before returning to Earth. But *Salyut 3* was doomed. The next crew found it difficult to rendezvous with the space station and used up too much fuel trying to dock with it. They returned to Earth, and the station was left unmanned until it was deliberately burned up in the atmosphere in January 1975.

ABOVE V. Bykovsky and S. Jaehn pictured inscribing their autographs on the sides of the landing ship of the Soyuz program on September 3, 1978.

RIGHT Launch of the Soviet *Soyuz 13* on December 18, 1973.

Salyut 4 suffered similar problems after its launch in December 1974. The first crew boarded the craft and stayed for a month. The second crew ran into trouble during the ascent, and safely returned to the ground never reaching orbit. The third crew had better luck and stayed aboard for two months. After these three missions, the station was brought down over the Pacific in February 1977. *Salyut 5* followed the same fate and was also unsuccessful.

At first, it seemed that *Salyut 6*, launched on September 29, 1977, would follow this dismal pattern. The first cosmonauts to arrive were unable to dock with the station, as its latching mechanism failed to operate. Though *Salyut 6* had a docking port at both ends, the crew had used up too much fuel and had to return to Earth.

The Russians took action to rescue the project. Space engineer Georgy Grechko was sent up in another Soyuz, assisted by Yuri Romanenko, on December 11, 1977. After reaching the station, they docked successfully at the second port. After unloading their equipment they entered *Salyut 6,* and prepared to examine the docking mechanism at the opposite end of the craft. To get there, Grechko had to climb out through an air lock and work his way around the outside surface using a series of handholds to reach the docking port. Romanenko also emerged from the air lock and waited to see if his help was needed.

To Grechko's surprise, the mechanism worked perfectly, so the problem must have resulted from a flaw with their predecessors' Soyuz. As Grechko was

BELOW The carrier rocket with the spacecraft Soyuz, designed by S.P. Korolyov, on show in Baikonur, Russia, January 1987.

relaxing and marveling at the spectacle of Earth, he was horrified to see Romanenko had lost his footing. He was drifting helplessly out into space, almost beyond reach. Grechko stretched out as far as he could, and managed to grasp the end of his comrade's disconnected lifeline. Carefully he pulled him back, and they climbed back through the air lock, only to find that the instruments were reporting an air leak.

This leak could have been deadly. Fortunately, it proved to be another instrumentation fault. From then on, all went smoothly. Grechko and Romanenko stayed in orbit for more than three months. When their mission was over, they were replaced by a whole succession of changeover crews, and by 1982 missions were lasting more than six months at a time. The space station concept, it appeared, was clearly here to stay.

BELOW An artist's concept of the mission profile of the Apollo-Soyuz Test Project, including launch, rendezvous, docking, separation and splashdown.

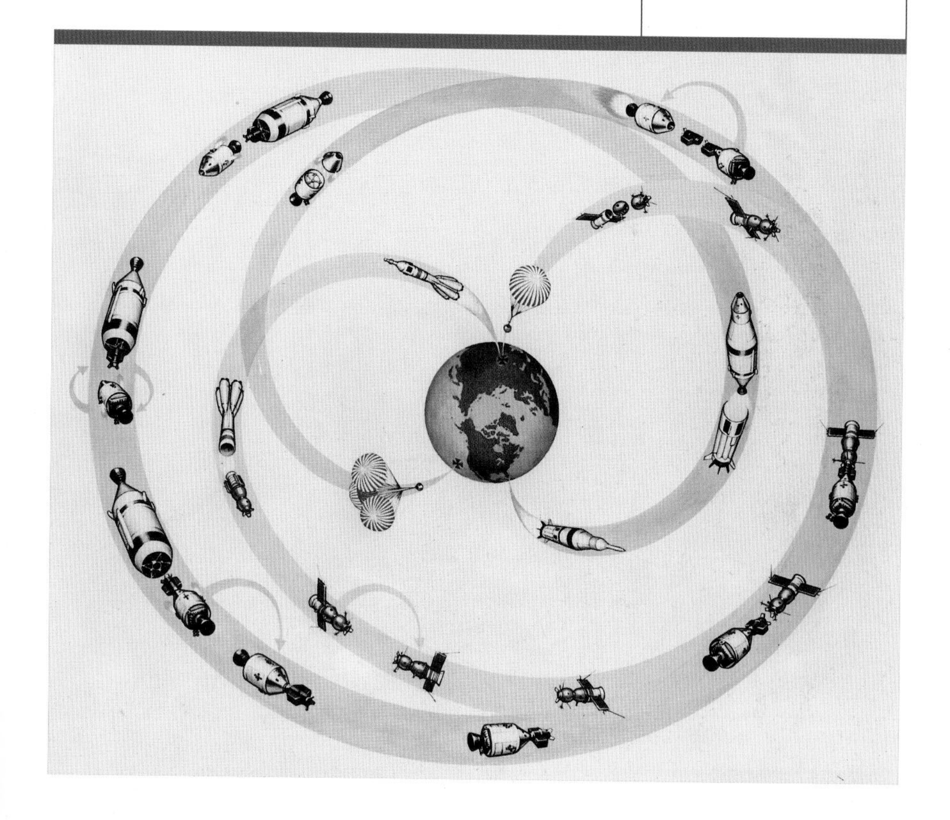

Emergency in space:
THE MIR STORY

The Russians embarked on their most ambitious space station program when they launched the core module of the Mir station on February 20, 1986. In addition to regular resupply and crew replacement craft, Mir had at least six docking ports, for specialized living and working modules to be attached to it.

In March 1987, the Kvant 1 astrophysics laboratory almost doubled Mir's size. Kvant 2 was added in February 1990, with washing facilities and an air lock for work outside the station. A month later, Kristall provided an additional docking port and more laboratory space. The Spektr module of March 1995 carried additional solar arrays and scientific equipment, while in April 1996, the Priroda module added more sensors and scanners.

The complete station weighed 130 tons and was used for 29 Soyuz flights and eight flights by the space shuttle, carrying around 100 different astronauts from various countries over its 13 years in space. Mir burst into prominence on June 25, 1997, when an

BELOW The damage done to the solar array panel is clearly shown on this view of Mir.

ABOVE A close-up view of the solar array panel on Mir, shows damage incurred by the impact of a Russian unmanned Progress resupply ship.

unmanned Progress supply vessel collided with the Spektr module while using a new guidance system for docking. The module lost pressure and electrical power, and the astronauts had to evacuate it and seal it off from the rest of the station. Working together during this emergency, Soviet cosmonauts and US astronauts tried to save the station rather than abandon it. Through their quick and resourceful actions, the crew and the station survived.

In all, Mir survived a total of 1,600 breakdowns. By the time the last crew left the station on August 27, 1999, 16,500 different scientific experiments had been conducted on board. In March 2001, Mir was brought back to Earth, where it crumbled and burned before plunging into the South Pacific.

8

TO THE SOLAR SYSTEM AND BEYOND

To the Moon, Mars, Venus and Mercury — unmanned probes can soar far beyond the range of human astronauts. How do they work, when were they launched and what have we learned from their travels?

BELOW An illustration of the Pioneer Jupiter spacecraft.

Almost from the start, space exploration has been pulled in two different directions — manned and unmanned. The spectacle of astronauts standing on the Moon, or on space walks aboard orbiting spacecraft, are the most lasting images of the Space

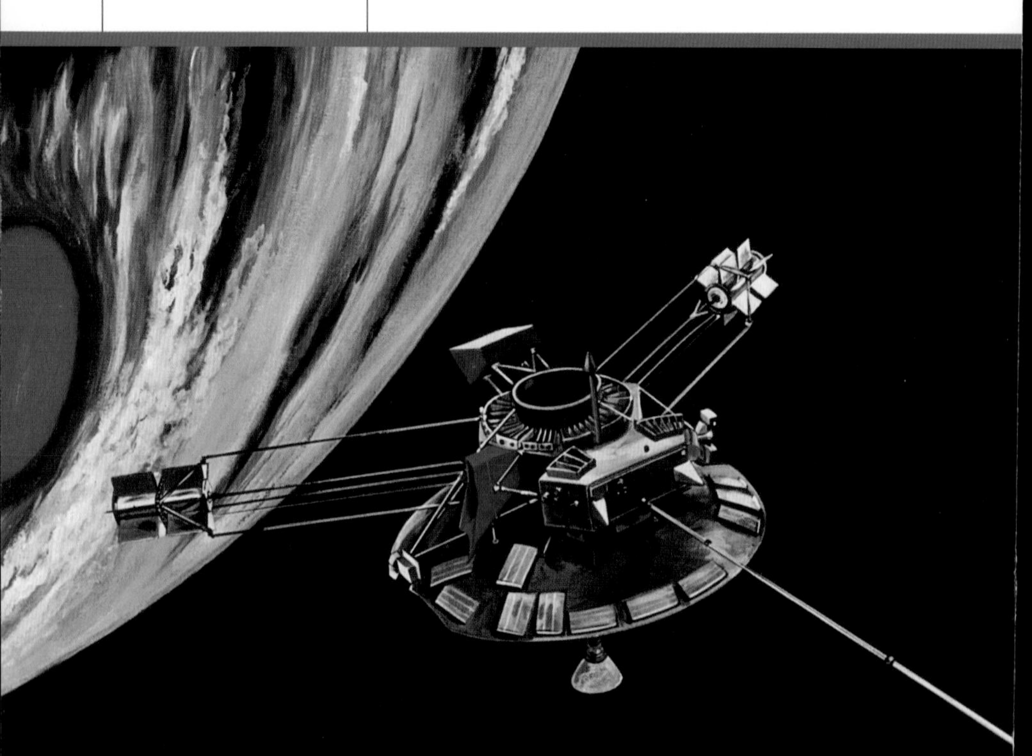

ABOVE Scientists working on the Ranger probe.

Age. But manned space ventures are costly and extremely complex. The huge budgets that enabled Neil Armstrong to take "one giant leap for mankind" could have sent a series of unmanned probes to places still far beyond the range of human astronauts.

The earliest spacecraft could not carry human crews. But the robot spacecraft soon began heading for more ambitious targets. On January 2, 1959, the Russians launched *Luna 1*. This probe was carried on the same type of rocket used for the early Sputnik Earth satellites, but was also fitted with an additional booster stage.

The power carried the 800-pound (363 kilogram) *Luna 1* into space, but navigation was a problem. The spacecraft missed its target, the Moon, by 3,700 miles (5,953 kilometers). Eventually, it took up an orbit around the Sun, sending back information on solar and cosmic radiation that allowed scientists to measure the density of matter in interstellar space. *Luna 2*, launched that same year in September, did score a direct hit on the Moon.

In those years, America's space program lagged behind the Soviet Union's. The first three Pioneer space probes did not complete their journeys. Like *Luna 1*, *Pioneer 4* missed the Moon by more than twice its designed separation of 15,000 miles (24,000 kilometers). Then it, too, went into orbit around the Sun. In the meantime, the Russians scored another coup with their *Luna 3*, launched on October 4, 1959, exactly two years after *Sputnik 1*. From an altitude of 4,900 miles (7,880 kilometers), *Luna 3* was able to

ABOVE The Surveyor spacecraft.

photograph almost three-quarters of the side of the Moon hidden from Earth observation.

Next, researchers worked on systems that could land a lunar probe safely. The lack of any atmosphere meant that a retro-rocket engine must be used to slow the craft. This required precise navigation and very sensitive control of the rocket thrust. While scientists worked on that problem, the remaining Ranger spacecraft were programmed to send pictures of the lunar surface up to the moment of impact, when the craft stopped working.

Once again the Russians triumphed, by making the first successful soft landing on another planet. Craft in their second Luna series were designed to land a 220-pound (100 kilogram) instrument package on the Moon, softly enough to function after the impact. The spacecraft made a single complete orbit of the Earth before the final stage rocket was fired to head the spacecraft on course for the Moon. Within 46 miles (74 kilometers) of its target, the radar and navigation equipment was dropped to reduce the landing weight. The retro-rocket engine then burned to reduce the descent speed. After the instrument package at the top of the probe was thrown clear, it was designed to lose energy by bouncing across the surface like a ball. Its heavily-weighted base caused it to stop right side up, allowing it to deploy its camera and antennae.

Even so, the equipment did not work well when the first eight probes were launched between 1963 and 1965. *Luna 9* finally succeeded in early 1966. For the first time, pictures were sent back to Earth from the lunar surface, over the four days the probe operated before its batteries ran down.

The Americans faced similar problems while designing their Surveyor probes to achieve a soft landing on the Moon. The retro-rocket was backed up by a set of vernier retro motors that started operating 52 miles (84 kilometers) above the surface. The retro-rocket engine was designed to burn out, and separate from the spacecraft at 37,000 feet (11,300 meters) above the surface. Then vernier motors would control the remainder of the descent. They would cut off at 14 feet (4.3 meters) from the surface, with the craft falling at just 3.5 miles per hour (5.6 kilometers per hour). The motors would then cut off to avoid disturbing the surface too much, and shock-absorbing landing pads would absorb the final fall.

When the first Explorer was launched four months after *Luna 9*, it hit its target first time. The equipment operated as planned, and the probe landed softly. Over a six-week period, it sent more than 11,000 pictures back to Earth. Two later Surveyors were lost, but others sent back thousands more pictures and data on the density and composition of the lunar surface.

Once they could make soft landings, both the Americans and the Russians sent ever more ambitious probes to explore different parts of the Moon. In September 1970, the Russian *Luna 16*

BELOW An illustration of the Surveyor probe touching down.

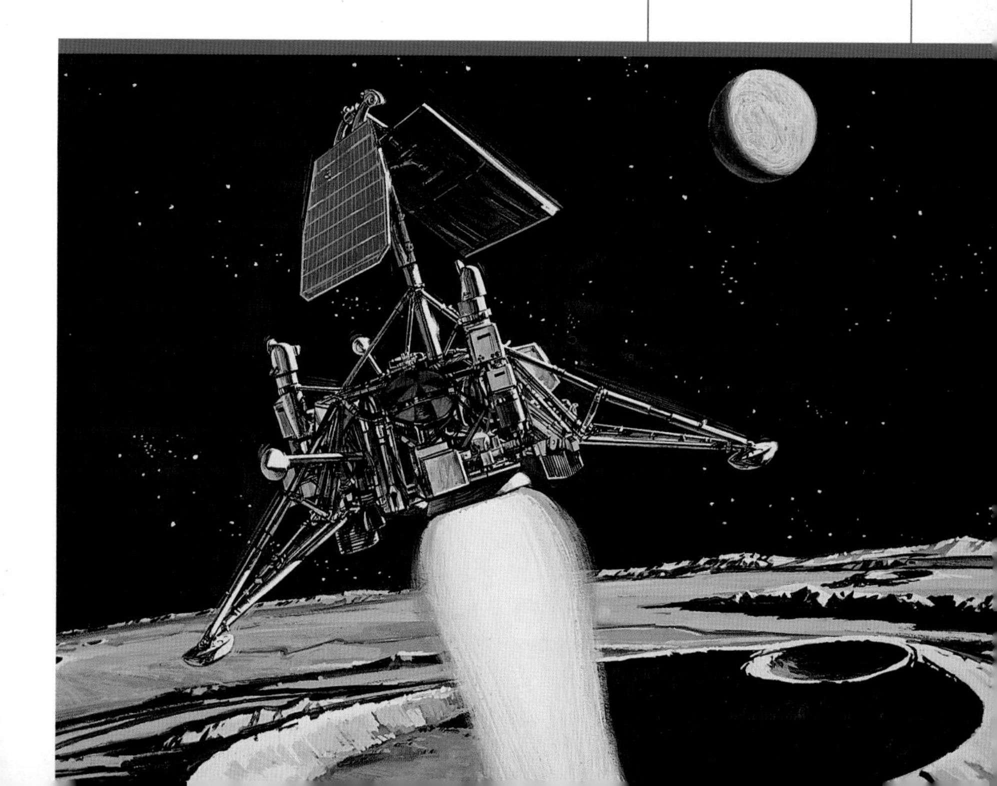

TOURING THE FARTHER REACHES

Imagine visiting most of the other planets in the solar system in just one voyage. The precise navigation techniques used to steer *Mariner 10* also enabled two US space probes to achieve that feat. *Pioneer 10*, launched in 1972, photographed Jupiter, the largest planet, and its satellites before heading past Pluto on its way into deep space. It sent more than 300 pictures, including images of the Great Red Spot, a huge atmospheric storm that humans first observed more than three centuries ago. Next came *Pioneer 11*, which was accelerated by Jupiter's gravity onto a trajectory for Saturn. At the end of six years, it passed within 21,000 miles (33,790 kilometers) of Saturn's ring system, before it, too, headed for the depths of space.

BELOW A montage of images of the Saturnian systems prepared from an assemblage of images taken by the *Voyager 1* spacecraft in November 1980.

Voyager 1 and *Voyager 2* were even larger and more sophisticated. Launched in autumn 1977, they reached Jupiter two years later. Both craft sent back high-quality pictures of the storms that racked the clouds covering Jupiter, and of the four largest of its 16 moons. Just over a year later, both Voyagers passed Saturn. They revealed a total of seven different rings of ice particles around the planet, and at least 22 moons, the smallest of which are only a few miles across.

Voyager 2 was then taken past Uranus, which it reached nine years after launching. Out there, in the darkest and coldest reaches of the solar system, Uranus appeared as a smooth, cloud-covered sphere, banded by 11 rings and circled by 15 satellites. Finally, after 12 years of space travel, the craft reached Neptune,

ABOVE This image of Saturn was taken 11 million miles (18 million kilometers) away by *Voyager 1*.

LEFT A view of the Voyager spacecraft full scale model.

revealing a cloud-covered planet with a faint ring system, eight moons, and an atmosphere of hydrogen, helium and methane.

Like their Pioneer predecessors, both Voyagers are heading out into deep space on very different trajectories. *Pioneer 10* will reach the vicinity of the star Aldebaran, but its journey will take eight million years. *Voyager 2* is heading for the star Sirius, which it will pass in a comparatively "brief" 358,000 years. By then, all the probes will long since have stopped radiating signals. But NASA scientists hope they can continue mining information, from the Voyagers in particular, for the next 30 years.

Some unmanned probes inspect other targets. When Halley's Comet returned in 1986, Russia, Japan and the European Space Agency launched probes to study its structure and composition. The Japanese Sakigake and Suisei probes passed ahead of its nucleus, while the Russians' *Vega 1* and *Vega 2* followed a trajectory that would take them within 5,600 miles (9,000 kilometers) of the nucleus. The European Space Agency's Giotto craft came closest of all — within just 375 miles (603 kilometers) of the comet's nucleus. Although the constant battering of fragments from the comet damaged its camera, the other sensors kept working. These probes showed that the comet's dark, dumbbell shaped nucleus was only a few miles across, but weighed 100 billion tons. It ejected jets of water and carbon monoxide and dust in its wake, as it traced its brilliant arc across the darkness of space.

ABOVE A view of Saturn's rings taken by *Voyager 1* on November 12, 1980.

landed in an area called the Sea of Fertility. Ground control commanded it to lower a drilling rig, which bored into the surface and lifted a soil sample that was stored in a chamber at the top of the craft. Twenty-six hours later, this section was brought back to Earth, carrying the soil sample. The lower section remained on the Moon, in order to measure temperature and radiation levels and transmit that data to Earth.

Two months later, *Luna 17* landed on the Sea of Rains and unloaded an eight-wheeled vehicle that could inspect the area. This first lunar rover, *Lunokhod 1,* covered more than 6 miles (9.5 kilometers) in a period of over 10 months, testing soil samples and sending back video pictures. In January 1973, *Lunokhod 2* covered more than 20 miles (32 kilometers) in the Sea of Serenity.

Next, engineers focused on the Sun. *Pioneer 5*, the first truly interplanetary probe, was launched on March 11, 1960, followed in 1965 by a series of later Pioneer probes. They were put into orbits around the Sun at both greater and lesser distances than that of Earth. These probes monitored the solar wind and its effects on Earth's magnetic field, among a variety of other things.

Meanwhile, the Russians had turned their attention to Venus. After two probes failed in 1961, in October 1967, the *Venera 3* sent back useful information from this mysterious planet. The probe reported atmospheric temperatures over 518 degrees Fahrenheit (270 degrees Celsius) and pressures 20 times higher than those on the Earth's surface, before the transmitter failed still 15 miles (24 kilometers) above the surface. Later, *Venera 7* finally revealed conditions at ground level on Venus. This amazing probe lasted fifteen minutes at a surface temperature of 891 degrees Fahrenheit (477 degrees Celsius), high enough to melt lead, and a pressure of 90 atmospheres.

In 1975 *Venera 9* and *10* went into orbit around Venus and each one released a probe that carried

OPPOSITE Lunar orbitor with a view of Earth in the background.

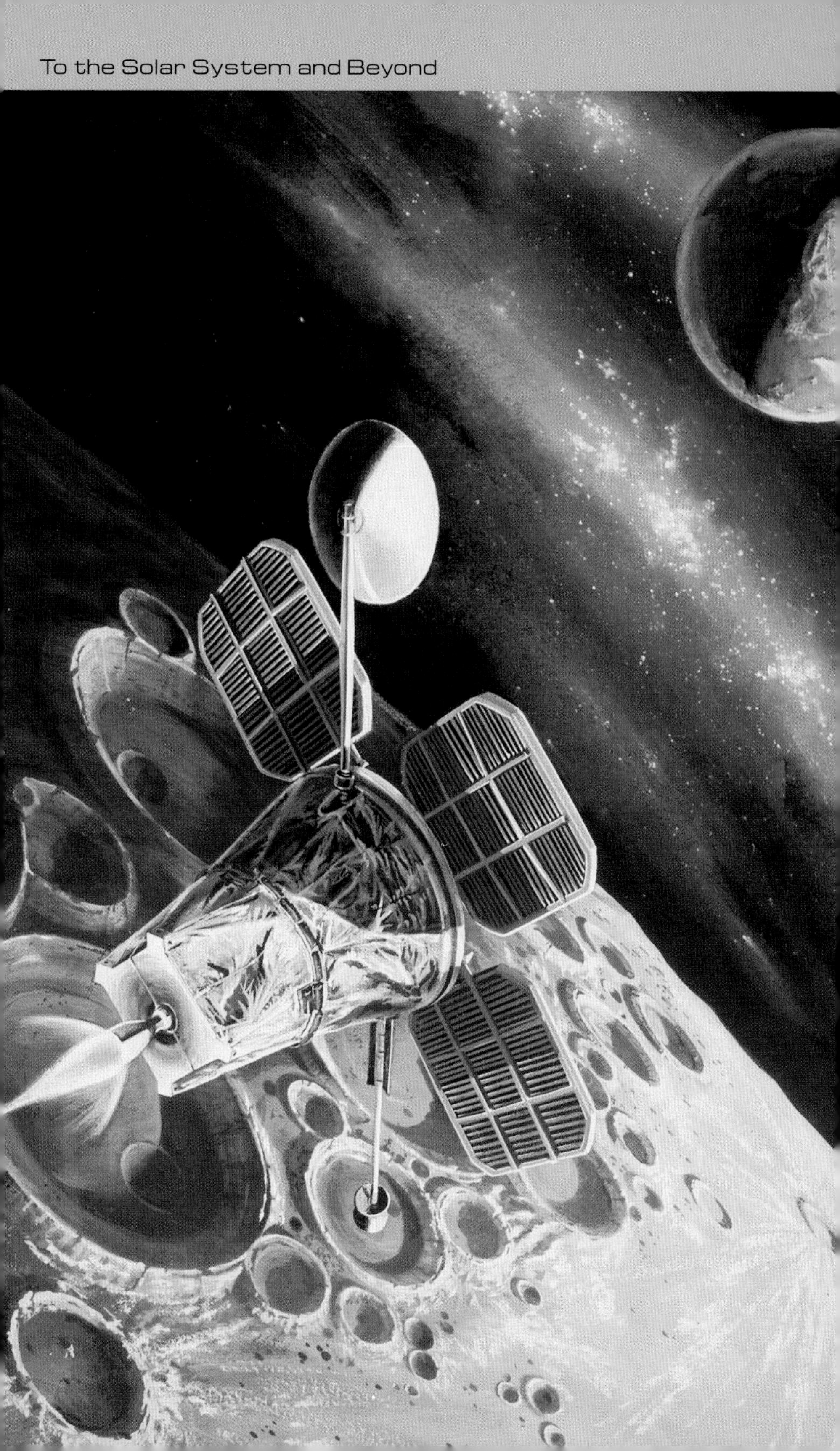

Hitching a ride to Mercury

***Mariner 10*, the American space probe launched in November 1973, had two goals. After the probe passed Venus and sent back the first video pictures of the planet, it followed a trajectory that used the gravity of Venus to change its course and slow it down. Then it performed a series of flybys over Mercury, the innermost planet of the solar system.**

The *Mariner 10* made three sweeps past Mercury, about six months apart, revealing an even more hostile world than either Mars or Venus. The atmosphere contains traces of argon, neon and helium, at a pressure one-trillionth that of Earth, and the barren surface is covered with impact craters. Its temperature varies between 210 degrees Fahrenheit below zero (–134 degrees Celsius) at night to a peak of 940 degrees Fahrenheit (504 degrees Celsius) under the Sun's fierce glow. At its closest approach, the Sun is 28 million miles (45 million kilometers) away, compared with the 93 million miles (150 million kilometers) that separate Earth from the Sun.

video cameras. The lander was designed to descend rapidly through the scorching atmosphere, with a parachute that released it into free fall while still 30 miles (48 kilometers) above the surface. Both landers sent back pictures revealing a barren stony surface. Later probes showed an atmosphere that was mainly carbon dioxide and clouds producing sulphuric acid rain, which boiled before it reached the surface, and endless violent thunderstorms. American probes added details and showed that Venus is more perfectly spherical than Earth, with generally flatter surface contours.

Robot explorers have reached Earth's other near neighbor, Mars, more easily. The American probe *Mariner 4* flew past the planet on July 14, 1965, at a range of 6,000 miles (9,650 kilometers). It sent back

OPPOSITE **A scientist works on the *Mariner 2* probe.**

ABOVE RIGHT ***Mariner 10* view of Mercury.**

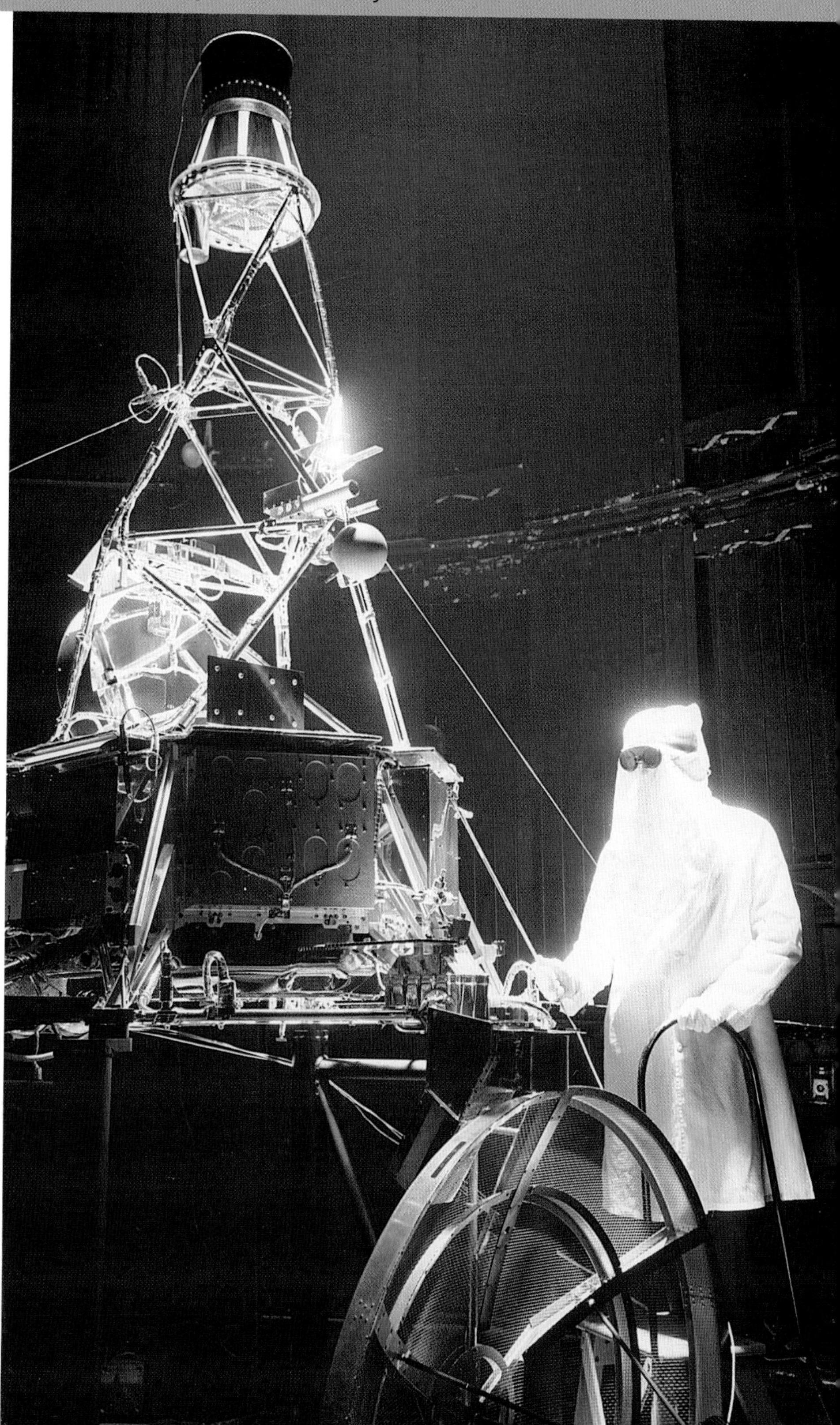

data showing the Martian atmosphere exerted only 0.01 atmospheres at the surface, which was covered in impact craters resembling those on the Moon. The distance separating Mars from Earth varies due to Mars's more elliptical orbit so NASA was able to send a much heavier and more complex spacecraft in 1971, when this distance approached its minimum.

When *Mariner 9* was placed in orbit around Mars, a vast dust storm was raging there, so large that it could be made out through powerful telescopes from Earth. Once the dust settled, the probe sent back more than 7,000 video pictures showing a series of huge dormant volcanoes. Some pictures showed a vastly deep rift valley, the Valles Marinaris, extending one-fifth of the way around the planet's surface. Networks of what appear to be long dried-up waterways, such as streams and rivers, hint that Mars once possessed abundant moisture.

In 1975, the American probes *Viking 1* and *Viking 2* topped this feat. *Viking 1* established itself in orbit around Mars, sending back video pictures of possible landing sites for the robot probe on board. Mission Control staff then maneuvered the spacecraft

RIGHT The Viking lander is shown, in an illustration, on the surface of Mars.

ABOVE ***Mariner 5*** **on its mission to Venus.**

to find an ideal site, and commanded the release of the lander vehicle over an area called Chryse Planitia. *Viking 1*'s lander touched down on July 20, 1976, followed by *Viking 2*'s lander, which surfaced on Utopia Planitia, on September 3. Both craft carried soil scoops and an analysis lab to investigate any signs of life. As they continued to send back data into the early 1980s, the signs were promising but uncertain. So the search continues, using ever more clever unmanned craft to look farther and farther.

Unmanned spacecraft were the first human-made objects to be launched into space, and still form the basis of space research, thanks to spin-offs from programs such as Skylab and Apollo. Today's robot explorers can land softly in harsh and distant conditions. The advanced navigational and control systems that carried astronauts into space have also steered cleverly designed probes across the farthest reaches of the solar system. Despite a lack of funds, more probes will be launched in the future. They will continue to explore space through the window that *Sputnik 1* opened over 40 years ago.

9

SPACE SPIN-OFFS

Computers, microchips, freeze-dried food, satellite communications, weather study and prediction, orbital telescopes, virtual reality programs and electric vehicles — these are just some of the useful products resulting from the space program.

Public interest for space exploration began to wane after the ultimate goal of placing humans on the Moon was achieved. Even the amazing feats of robot probes and landers, and the missions of the space shuttle, failed to arouse interest in space research.

As for the Russians, the breakup of the Soviet Union in 1989, and the country's crippling economic problems, cramped its once-thriving space programs. Once Russia withdrew from the East-West space

RIGHT The Wind spacecraft, part of the global Geospace Science Initiative to measure properties of the solar wind before it reaches earth.

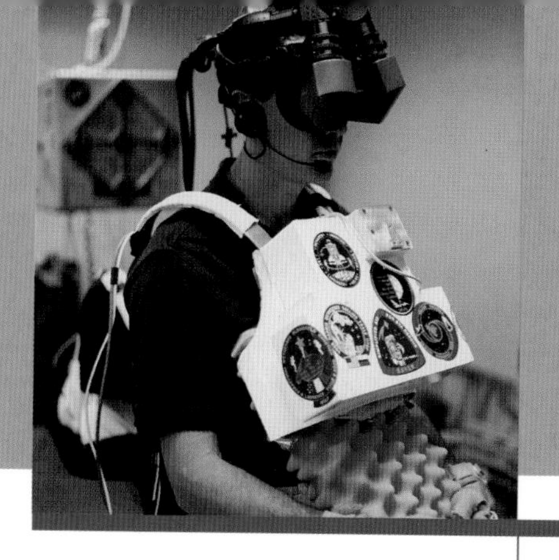

ABOVE Astronaut James H. Newman undergoing virtual reality training before his mission to the *International Space Station.*

race, America had no urgent political reason to compete. To maintain its funding and continue its work, NASA had to convince an increasingly uninterested public that space exploration produces many practical benefits.

Obviously, the Apollo program required a huge budget in order to protect the astronauts' lives. Humans traveling in space and landing in an airless, waterless world could die within minutes if anything went wrong. Every detail had to be considered and safety could not be compromised, regardless of cost.

Did those vast costs bring many practical benefits in return? The answer is quite reassuring. The current space program accounts for less than 1 percent of the American budget. On average, every dollar spent on the space program returns $7 to the economy, from increased jobs and economic growth.

But the biggest gain of all cannot even be measured. The long list of spin-offs from space technology into our daily lives is varied, and still growing. One recent addition to the list is rescue blankets made from recycled plastic milk bottles that can keep people warm, even when wet. This idea sprang from NASA's research into lightweight metal insulation for spacecraft. The CCD chip technology used in the *Hubble Space Telescope* is also being used in a device for imaging breast tissue more clearly in order to spot the minute differences between benign and malignant (cancerous) tumors.

The Apollo moon program required much more powerful yet smaller computers, which must fit into packages as light and compact as possible. This stimulated the creation of the microchip. All the separate elements of an electrical circuit such as resistors, capacitors and transistors were combined into one precisely engineered wafer of semi-conductor material. Microchips were also cheaper and faster to produce.

Sometimes the links between space technology and other products are less direct. For instance, NASA research into food suitable for long periods in space led to the creation of a vegetable oil substitute based on microalgae. This oil contains two fatty acids found in human milk but missing from most baby formulas, so it can be used to develop enriched baby foods. Likewise, the technology developed to sterilize water in space is now being used to purify swimming pools, without chemicals: passing a current through silver-copper alloy electrodes produces silver and copper ions that kill bacteria.

The thermoelectric technology used to cool spacecraft has also made it possible to deliver large amounts of heating or cooling capacity from low-power sources, such as car cigarette-lighter sockets. Plug-in accessories can heat water to 125 degrees Fahrenheit (52 degrees Celsius) or provide the cooling power of a 10-pound (4.5 kilogram) block of ice. The environmental control system made for the space shuttle has also been used to produce the Barorator. This device monitors changes in atmospheric pressure and calculates the immediate rate of change as an aid to forecasting the weather.

Other successful spin-offs include personal alarms, emergency rescue cutters, storm warning systems, and better brakes and tires. Corporate jet aircraft are flying with more efficient wing sections using NASA computer programs. Space technology even extends into articles like wheelchairs, school buses and television screens. Though people may not realize that space research produced these benefits, future space exploration will continue to change our lives in many different ways.

OPPOSITE The giant *Hubble Space Telescope* that was useful in the development of CCD chip technology.

SPACE MEDICINE

and the weightless factory

Applied space technology has brought important new tools for health care. For instance, women at high risk for breast cancer are routinely given X-ray examinations. The use of a solar cell sensor, placed directly below the X-ray film, allows the system to be turned off automatically when the film has been exposed to enough radiation to provide a clear picture. This speeds up the exam so that more patients can be treated, and it reduces radiation hazards.

NASA ultrasonics technology is increasingly being used to find out the depth of a burn victim's injuries. That way, doctors can match the treatment more precisely to the actual damage. Furthermore, image processing techniques first developed for the space program are now used to detect eye problems in very young children.

Robot technology has been used to make a patient-controlled wheelchair that can respond to 35 one-word voice commands, and modified space suit designs have produced a cool suit used to treat people with multiple sclerosis, spina bifida, cerebral palsy and other conditions. Perhaps most dramatically of all, the sensor and control systems used in satellites have been harnessed to help victims of chronic pain or involuntary movement disorders, with an implant device that electrically stimulates certain nerve centers, or certain areas of the brain.

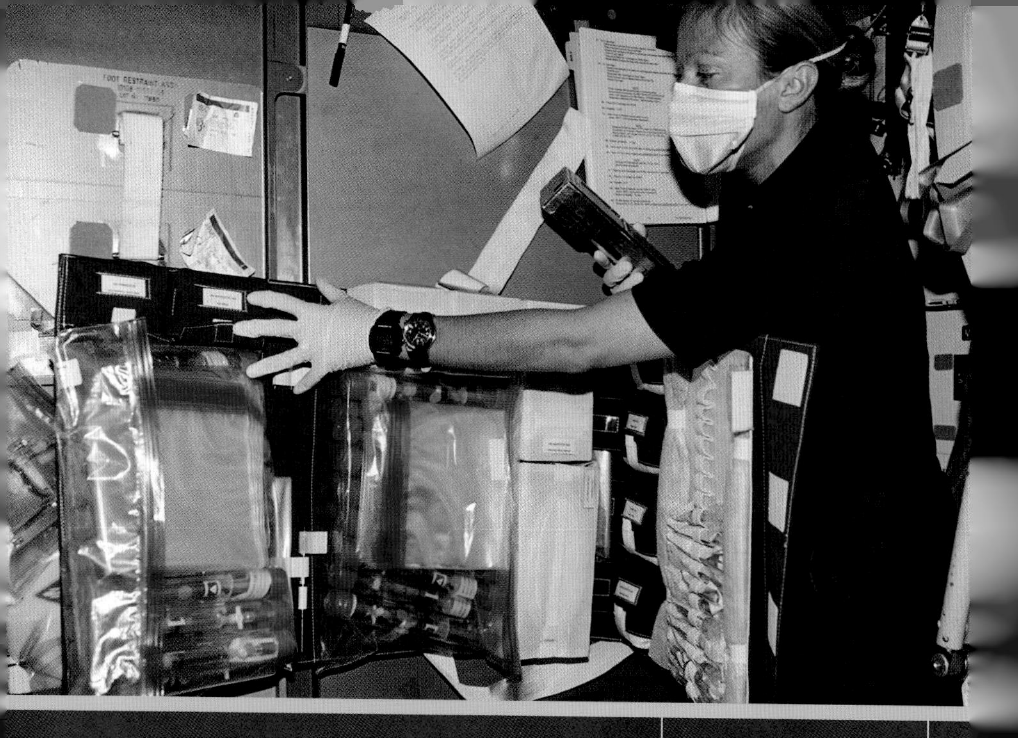

Drugs have also been purified in space. Using electrophoresis equipment specially designed to make the most of the weightless conditions can produce greater yields and higher purity than can be made with earthbound equipment.

Space technology is used to improve ground-based industrial processes, or machinery, too. Magnetic bearings that were developed for the Space Shuttle are also used to support moving machinery without the need for direct physical contact. This avoids losses from friction, wear of the bearing surfaces or unwanted heat production, which helps in the operation of machine tools.

Even the production of spacecraft such as the Shuttle has spawned technologies that can aid other industries. The need for high-quality welds to insure the integrity of the external fuel tank produced a superior automated laser welding system. This is now being used in industry where similar high-quality welding is needed.

ABOVE Astronaut Kathryn P. Hire works in the neurolab on board the shuttle *Columbia.*

OPPOSITE Jay C. Buckey prepares to conduct a neurolab sleep experiment.

10

AN EXCITING FUTURE

Take a glimpse at what lies ahead in space exploration: colonies on the Moon, flights to Mars, deep space probes and nuclear fusion rockets pushing us at awesome speed into the deep unknown.

ABOVE An advanced artist's concept of the *Near Earth Asteroid Rendezvous* spacecraft rendezvousing with the asteroid Eros to send back data to Earth.

Following decades of stunning achievements in space, future goals are less clear. The Soviet Union has virtually dropped out of the space race. While international cooperation will aid future projects, the pressure to beat the opposition to the next milestone is gone. This competition once spurred both the Americans and the Russians to gain money and support for new research. Fortunately, most of the key technology for space exploration was already in place by the 1990s.

Basically, there are four options. Launching more sophisticated satellites and space probes would tell us more about the solar system and the wider universe. The technology already exists to build more ambitious space stations. These can provide bases for scientific research, and for long-distance space travel, in case funds become available. With present technology, it would be feasible to set up a long-term colony on the Moon, for a new view of the universe, and a source of the materials needed for life on Earth and for long-distance space flight ... or to place human astronauts on Mars.

Despite the exceptional planetary probes launched in recent decades, our knowledge remains limited. Only one face of Mercury has been filmed. We know very little about the surface of Venus or the polar regions of Mars. The outermost planet, Pluto, with its single moon, remains mysterious. During the first

ABOVE This illustration shows a crew landing on Mars and using an unpressurized rover to unload cargo and supplies for their stay.

years of the 21st century, Pluto is heading out away from the Sun, along its long and unusual orbit. Yet while its thin atmosphere seems to be cooling since the 1980s, there are signs that Pluto's surface is becoming slightly warmer.

In 2000, NASA cancelled a planned mission to Pluto due to funding shortages. Supporters of this mission pointed out that unless a probe reached Pluto by at least 2020, the chance to study this tiny planet's atmosphere would be lost for 230 years. In December, NASA said that it might reconsider its decision and began looking at proposals to complete the project. The goal was to reach Pluto by 2015, which would mean launching a probe no later than 2006. In July 2002, a Senate subcommittee approved funding $115 million for the project, but its future was, sadly, unclear.

The development of smaller, cheaper and more reliable spacecraft could make the Pluto launch possible. Already the first of several new, low-cost NASA probes is exploring the mysterious asteroid belt between the orbits of Mars and Jupiter. The *NEAR (Near Earth Asteroid Rendezvous)* probe, launched by a Delta rocket in 1996, headed for the asteroid Eros, measuring about 15 miles by 9 miles by 9 miles (24 by 14.5 by 14.5 kilometers), and began orbiting the asteroid in February 2000.

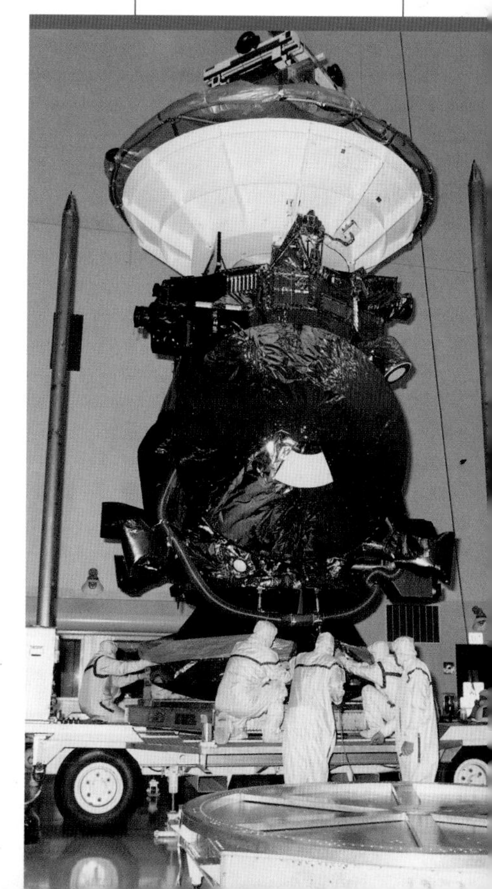

BELOW The *Cassini* spacecraft, which is scheduled to go on a four-year close-up mission to study the Saturnian system.

LOOKING FOR LIFE

Whether aliens exist or not, it remains impossible to send manned spacecraft over the kinds of distances needed to make contact. Scientists have found ways to send messages, independent of language, on unmanned deep space probes. These messages would show any life-forms encountering them something about the people who created and sent them.

This was first done in the early 1970s with the *Pioneer 10* and *11* spacecraft, sent to pass Jupiter before heading out into deep space (see chapter eight). Both of these craft carried plaques showing outlines of male and female human figures. They also showed the position of Earth relative to the Sun and showed the sun's position relative to important radio stars. These craft will keep traveling through space, possibly for millions of years, unless captured by the gravity of some massive star or planet. So, it is vaguely possible that they may eventually be intercepted by aliens.

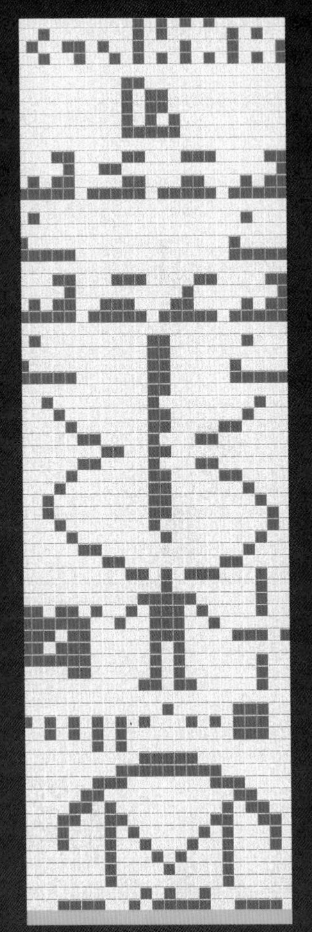

A much faster way to relay information across space is to send it as powerful radio signals traveling at the speed of light. In 1974, a series of 1,679 pulses was sent from the huge Arecibo radio telescope in Puerto Rico. These pulses could be arranged to assemble a diagram showing a human figure, information on prime numbers, the structure of the solar system and the radio telescope sending the signals.

The messages were beamed out in the direction of a cluster of 300,000 stars in the Hercules constellation. These are 24,000 light years away, so this message will take 24,000 years to arrive. And any reply would reach Earth a full 48,000 years from now!

The search for extraterrestrial signs of life is one of the most exciting goals of future space missions. Organic molecules have been found in meteorite

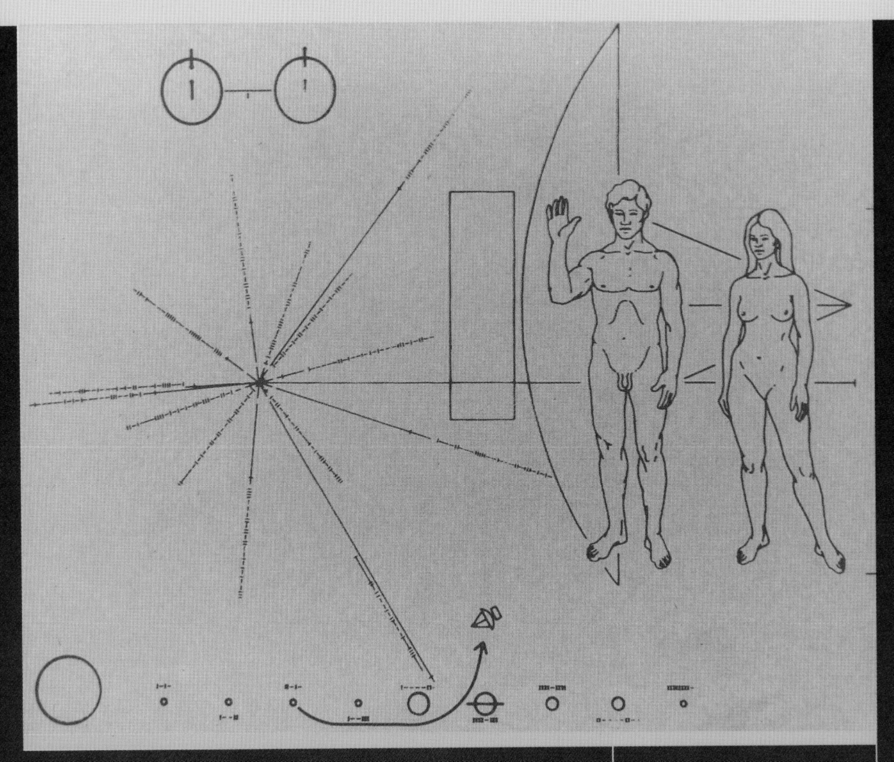

ABOVE The Pioneer plaque, attached to the spacecraft's antenna.

OPPOSITE A pictogram showing the information coded and sent by radio telescope into outer space.

fragments. Besides, the discovery of planetary systems on neighboring stars like Upsilon Andromeda that may include planets similar in size, composition and temperature to Earth, are signs that life might exist beyond our planet.

In the shorter term, Mars is the most promising place to start the search. As long ago as 1988, the Soviet Union announced plans to send two nuclear-electric powered spaceships to Mars, where they would begin to orbit the planet. Then cosmonauts would descend to the surface aboard specially developed modules that would also carry roving vehicles, with the second ship available for rescue or resupply. Humans would be able to cover a much wider area in their search for life than even the most advanced automatic lander.

Another possible approach would involve landing on the Martian moon Phobos and establishing a scientific base there. An unmanned vehicle would be

ABOVE An artist's concept of a future outpost on Mars.

launched first, to carry the return vehicle and the expedition's equipment, and to maintain it in orbit around Mars. A month later a second ship carrying four astronauts would follow. They would dock with the first spacecraft. Then two astronauts would descend to the surface of Phobos, in order to set up the long-term monitoring experiments needed to watch for the requirements of life.

In theory, these bold projects are possible, but three main problems must be solved. The first is financial, since the cost would probably be much higher than the sums spent on the Apollo missions. Another is biological, since astronauts would have to spend well over a year in zero-gravity conditions to cover the huge distances involved. The third is technological, since the different systems would have to function with a high level of reliability over longer distances.

Will these barriers be overcome? The possibilities are exciting. While we have learned a lot about space, humans still have many unanswered questions as we continue the ancient quest to understand outer space.

INTERNATIONAL SPACE STATION

The *International Space Station* will be the most complex and ambitious structure ever made in space and will include the *Columbus* laboratory module developed by the European Space Agency. In all, 100,000 people are working on different parts of the ISS, and 16 different nations are involved in this 10-year program, which began in 1998: the United States, Russia, Belgium, Brazil, Canada, Denmark, France, Germany, Holland, Italy, Japan, Norway, Spain, Sweden, Switzerland and the UK. The complete structure will weigh 460 tons and be completed by 2004, with 33 American shuttle flights and the launching of 12 Russian rockets. It will orbit Earth at an altitude of some 280 miles (950 kilometers) and a speed of 18,000 miles an hour (28,960 kilometers an hour), and its footprint will cover 85 percent of Earth's surface. A total of 52 computers will monitor its operations.

The station will have room for six or seven occupants. There will be two storage modules and seven laboratories and a robot arm, larger than the one fitted to the space shuttle. Over the 10-year planned life of the station, a total of some 900 scientists will be able to study space and conduct medical and material experiments.

BELOW The connecting of the Russian supply module *Zarya* with *Unity*.

Air lock an airtight chamber located between areas of unequal air pressure.

Asteroid a small celestial body that orbits the sun.

Constellation an arrangement of stars that stands out in the sky and has been given a name.

Dehydrated deprived of water.

Docking the process of coupling two spacecraft in space.

Docking adaptor a device that is used to provide a stable docking point between two space vehicles.

Elliptical orbit a path that moves in the shape of a curve that is formed by the intersection of a circular cone and a plane cutting through the cone.

Galaxy a large-scale collection of stars, gas and dust that contains an average of 100 billion solar masses and ranges in diameter from 1,500 to 300,000 light years.

Geosynchronous orbit an orbit in which a satellite revolves around the Earth at the same rate as the Earth rotates so that it always stays in the same spot in the sky.

Gravity the force of attraction which bodies with mass have for each other.

Heresy an idea or doctrine that conflicts with the established religious beliefs.

Interstellar between the stars.

Jettison to cast off or discard.

Nova a star that suddenly becomes much brighter, then gradually returns to its original appearance. (see also Supernova).

Nuclear reaction a reaction that changes the energy, composition, or structure of an atomic nucleus.

Orbiter the section of the space shuttle that is placed in orbit.

Organic relating to or derived from a living organism.

Payload the total load, in terms of cargo and/or passengers that a spacecraft can carry.

Probe an unmanned spacecraft that carries instruments to photograph, or obtain data in some other way, on bodies in the solar system or on interplanetary space.

Prototype an original form or model used as a basis for making later versions.

Rocket a device propelled by the ejection of combustion products of solid or liquid fuel.

Satellite an object built for the purpose of orbiting the Earth or another celestial body.

Sextant an instrument used in navigation to measure the altitudes of celestial bodies.

Solar array a device that converts energy from the sun into electrical power.

Solar system the sun and the planets and other celestial bodies that orbit the sun.

Space station a large manned artificial satellite used by astronauts as a base for scientific research in space.

Spacesuit a suit that is designed to allow astronauts to safely move outside their spacecraft.

Spectrum the distribution of energy given off by a radiant source arranged in order of wavelengths.

Spectral imaging the process of making images based on the spectra of objects or gases.

Suborbital flight a flight that involves less than one full orbit of the Earth.

Sunspot a dark spot that appears in groups on the surface of the sun.

Supernova a vast explosion of a huge star in which the star briefly becomes many hundreds of thousands of times brighter.

Telescope an instrument containing lenses and/or mirrors that gathers more light than the naked eye, enabling people to see faint objects in greater detail.

Thrust the forward-directed force that develops in a jet or rocket engine in response to the rear-ward ejection of fuel gases.

Trajectory the path of a moving body or particle.

Unmanned probe (see Probe)

Vacuum a space that is relatively empty of matter.

Velocity rapidity of motion; speed.

Zeppelin a rigid airship that is kept afloat by gas.

Index

Page numbers in italic type refer to illustrations.

Picture Credits

The publisher would like to thank the following for permission to reproduce images. While every effort has been made to ensure this listing is correct, the publisher apologizes for any omissions.

British Film Institute: 38. **Fortean:** 24. **Genesis:** back cover, 26, 30-31, 33, 36, 37, 42, 46, 51, 72, 101, 102, 103, 109, 111, 112, 113, 115. **NASA:** front cover, 1, 2, 5, 12, 13, 14-bl, 15, 16-t+b, 17-t+b, 18, 19, 20, 21, 25-bl, 39, 45, 52, 53, 55, 56, 62, 63-t+b, 64, 66, 67, 68, 69-b+r, 70, 71-t+b, 73-t+b, 76, 77-t+b, 78, 79, 80-tl, 81-t+b, 82, 83-t, m+b, 84, 85, 86, 87, 88, 89-t+b, 90, 91, 92, 93, 94-t+b, 97, 98, 99, 100, 104, 105-t+b, 106-107, 110, 114, 117, 118, 119, 120, 121-t+b, 122, 123, 124, 125. **Novosti:** 23, 34, 35, 38, 40-r+l, 41, 48, 50, 74, 75, 95-t+r, 96. **Science and Society Picture Library:** 6, 7-t+b, 8-t+b, 9-t+b, 10, 11, 14-tl, 22, 25-r, 27, 28, 43, 44, 47, 49, 57, 61, 80-bl.